Strengthening the Defense Innovation Ecosystem

BRODI KOTILA, JEFFREY A. DREZNER, ELIZABETH M. BARTELS, DEVON HILL, QUENTIN E. HODGSON, SHREYA S. HUILGOL, SHANE MANUEL, MICHAEL SIMPSON, JONATHAN P. WONG

Prepared for the Office of the Secretary of Defense
Approved for public release; distribution is unlimited

 RAND | NATIONAL DEFENSE RESEARCH INSTITUTE

For more information on this publication, visit **www.rand.org/t/RRA1352-1**.

About RAND

The RAND Corporation is a research organization that develops solutions to public policy challenges to help make communities throughout the world safer and more secure, healthier and more prosperous. RAND is nonprofit, nonpartisan, and committed to the public interest. To learn more about RAND, visit www.rand.org.

Research Integrity

Our mission to help improve policy and decisionmaking through research and analysis is enabled through our core values of quality and objectivity and our unwavering commitment to the highest level of integrity and ethical behavior. To help ensure our research and analysis are rigorous, objective, and nonpartisan, we subject our research publications to a robust and exacting quality-assurance process; avoid both the appearance and reality of financial and other conflicts of interest through staff training, project screening, and a policy of mandatory disclosure; and pursue transparency in our research engagements through our commitment to the open publication of our research findings and recommendations, disclosure of the source of funding of published research, and policies to ensure intellectual independence. For more information, visit www.rand.org/about/research-integrity.

RAND's publications do not necessarily reflect the opinions of its research clients and sponsors.

Published by the RAND Corporation, Santa Monica, Calif.
© 2023 RAND Corporation
RAND® is a registered trademark.

Library of Congress Control Number: 2022922748
ISBN: 978-1-9774-1024-5

Cover: Gorodenkoff/Getty Images/iStockphoto.

Limited Print and Electronic Distribution Rights

About This Research Report

This report documents findings and recommendations from a project examining how Department of Defense (DoD) innovation organizations can more effectively create and strengthen a "commercial technology pipeline" through which innovative commercial technologies can be identified, developed, and transitioned from the private sector to DoD for military use at both the joint and service level. The research team used a mixed-method research approach to gather and analyze data on what defense innovation organizations are seeking to do, the activities they conduct, and the outcomes they are seeking; and then identify the functions and other factors that are required to effectively accelerate the identification, development, and adoption of commercial technologies for military use. The research team identified challenges and gaps in current organizational structures and processes and developed and tested alternative approaches to strengthen these structures and processes. Based on this analysis, RAND's National Defense Research Institute (NDRI) developed recommendations to strengthen the commercial technology pipeline.

The summary of this report, which highlights our key findings and recommendations, will likely be most useful for decisionmakers. Readers with more detailed interests will find value in the chapters and appendixes, which provide detailed explanations of our methodology, study findings, and recommendations.

The research reported here was completed in September 2022, completed quality assurance, and underwent security review with the sponsor and the Defense Office of Prepublication and Security Review before public release.

RAND National Security Research Division

This research was sponsored by the National Security Innovation Network (NSIN) and conducted within the Acquisition and Technology Policy Center of the RAND National Security Research Division (NSRD), which operates NDRI, a federally funded research and development center sponsored by the Office of the Secretary of Defense, the Joint Staff, the Unified Combatant Commands, the Navy, the Marine Corps, the defense agencies, and the defense intelligence enterprise.

For more information on the RAND Acquisition and Technology Policy Center, see www.rand.org/nsrd/atp

Comments or questions on this report should be addressed to the project leaders, Jeffrey Drezner at zner@rand.org and Brodi Kotila at bkotila@rand.org.

Contents

About This Research Report iii
Boxes, Figures, and Tables vii
Summary ix
Acknowledgments xxi

CHAPTER 1
Introduction 1
    Focus of This Report 2
    Key Terms 2
    Structure of Report 3

CHAPTER 2
Approach 5
    Research Tasks 5

CHAPTER 3
The Commercial Technology Pipeline and Defense Innovation Organizations 15
    Developing a CTP Model 16
    CTP Activities, Functions, and Processes 19
    Utility of the CTP Model as a Framework 22
    Characteristics of a Well-Functioning CTP 22
    DIOs and the CTP 22
    General Findings Applying the CTP 33

CHAPTER 4
Challenges and Gaps 35
    Challenges and Gaps in CTP Characteristics 36
    Challenges and Gaps in CTP Functions 43
    Challenges: The Bottom Line 64

CHAPTER 5
Approaches to Strengthen the Commercial Technology Pipeline 65
    Identifying Alternative Approaches 65
    Approaches to Cultivate Desired CTP Characteristics 66
    Approaches to Strengthen CTP Identification, Development, and Adoption Functions 74

CHAPTER 6
Key Findings, Recommendations, and Considerations for Implementation................83
    Key Findings..83
    Recommendations to Strengthen the CTP 86
    Considerations for Implementation 92
    Further Research 94

APPENDIXES
A. **Defense Innovation Organization Deep Dives**...............................97
B. **Technology Case Studies** 141
C. **Acquisition Policy Game Design**...179

Abbreviations...185
References 189

Boxes, Figures, and Tables

Boxes

3.1.   Teeing Up Successful Transition 31
3.2.   Bridging the Valley of Death 32
5.1.   Foster a Shared Sense of Mission Across All CTP Stakeholders 67
5.2.   Develop and Promulgate Strategy, Plans, Policies, and Guidance for the CTP
        to Establish Common Goals, Objectives, and Outcomes 69
5.3.   Define and Communicate Roles and Responsibilities for the CTP, Ensuring
        There Are No Gaps in Key Functions 70
5.4.   Facilitate Information Sharing, Coordination, and Collaboration Across CTP
        Stakeholders 72
5.5.   Implement Incentive Structures, Including Metrics and Accountability
        Mechanisms, to Align CTP Stakeholders to CTP Goals, Objectives, and
        Desired Outcomes 73
5.6.   Develop More Rigorous, Comprehensive, and Shared Approaches to
        Technology Scouting 75
5.7.   Encourage CTP Stakeholders to Share Problems Across the Enterprise 76
5.8.   Establish a Single, Comprehensive, Integrated, Searchable Portal for
        New Entrants 78
5.9.   Establish Navigation Support Services to Help New Entrants Understand and
        Navigate DoD 79
5.10.   Assign Responsibility for Oversight of the CTP, Including Transition
        Outcomes, to an Organization with the Authority and Budget to Execute
        Those Responsibilities 80
5.11.   Provide Flexible Funding to Incentivize Collaboration and Facilitate
        Transition Gaps 81

Figures

S.1.   Representation of the Commercial Technology Pipeline xiii
3.1.   Representation of the Commercial Technology Pipeline 18
C.1.   Activities Available to Players 182

Tables

S.1.   Definitions of Key Terms x
S.2.   Challenges and Gaps in CTP Characteristics xv
S.3.   Challenges and Gaps in CTP Functions xvi
1.1.   Definitions of Key Terms 4
2.1.   Candidate Organizations for Deep-Dive Profiles 10

2.2.   Technology Case Studies 12
3.1.   Definitions of CTP Functions 20
3.2.   Characteristics of a Well-Functioning CTP.................................... 23
3.3.   Mapping Innovation Organizations to CTP Functions 24
4.1.   Challenges and Gaps in CTP Characteristics................................ 35
4.2.   Challenges and Gaps in CTP Functions.. 44
5.1.   Approaches Evaluated to Cultivate Desired CTP Characteristics ... 67
5.2.   Approaches Evaluated to Strengthen CTP Identification, Development, and
         Adoption Functions.. 74

Summary

Issue

Technological superiority is a key element of the U.S. military's strength and advantage over its competitors. While the government played a key role in sponsoring science and technology research during the Cold War, in recent years, technological innovation has been driven by the commercial market. Today, as China is aggressively modernizing and growing its military, the technological advantage the United States holds over its near-peer rivals has narrowed. Department of Defense (DoD) direct investment in basic research and development remains critically important but is not sufficient to retain the U.S. technological advantage. Rather, DoD must find a way to identify, support, and leverage innovation relevant to national security that occurs in the private sector, much of which is driven by the commercial marketplace and is conducted by individuals and entities that have traditionally not worked with DoD.

Recognizing this need, the Office of the Secretary of Defense (OSD) and the military services have created a number of defense innovation organizations (DIOs) in the last two decades to help foster communities of innovators and accelerate the identification, development, and adoption of private-sector-developed commercial technology into the military. Some observers have begun to ask how well these organizations have met their stated aims in promoting innovation and transitioning new commercial technologies to the warfighter, and whether the existing defense innovation ecosystem is optimally organized to support innovation and the transition of emerging commercial technologies for military use.

To address these issues, DoD's National Security Innovation Network (NSIN), a DIO that reports to the Under Secretary of Defense for Research and Engineering, requested the assistance of the RAND National Defense Research Institute (NDRI) in understanding how these innovation organizations can more effectively create and strengthen the activities, functions, and processes by which DoD identifies, develops, and transitions innovative commercial technologies from the private sector to DoD for military use at both the joint and service level. Taken together, we will refer to these activities, functions, and processes as the commercial technology pipeline (CTP). These and other key terms appear in Table S.1.

Key Terms

TABLE S.1
Definitions of Key Terms

Term	Definition
Defense innovation	The processes of generating and fielding technologies and other products, services, processes, or practices that are new or improved in the defense context
Defense innovation ecosystem	The set of organizations, activities, functions, and processes that develop, produce, and field new or improved technologies and capabilities for military use
Defense innovation organization (DIO), innovation organization	DoD organizations that were founded to help the U.S. military pursue innovative ways to sustain and advance the capabilities of the force, including making faster use of emerging commercial technologies
	Includes long-standing organizations such as the Defense Advanced Research Projects Agency (DARPA) as well as the new generation of innovation organizations created in the last 15 years, such as the Defense Innovation Unit (DIU) and AFWERX[a]
Commercial technologies	Technology originating from or matured by a commercial or private-sector entity
	Includes technologies or end products initially developed by the private sector (nondefense industry) without government funding and intended for the commercial marketplace
	Also includes technologies that were initially developed in DoD labs, licensed to the private sector to further develop and mature the technology or product through the commercial marketplace, and then sold back to DoD as a mature technology or product
Commercial technology pipeline (CTP)	The activities, functions, and processes by which DoD identifies, develops, and adopts commercial technologies for military use
CTP core function	Functions that occur within the identification, development, or adoption phases of the CTP
CTP enabling function	Support functions that occur within and across each of the three CTP phases
CTP phase	One of three stages in the CTP: identification, development, and adoption
CTP stakeholder	DoD entities that have an interest, concern, or role in the CTP
	Includes requirements and capability developers, acquisition organizations (program managers, program executive officers, system commands), budget offices, and end users at both the individual unit level as well as major and combatant commands
New entrant	A commercial or private-sector entity that has not previously sold to DoD
	Includes but is not limited to "early-stage ventures," which are new entities (business, person, or group) with a new idea or concept for technology or an application of technology
Transition	A shift in responsibility or ownership of the technology product or system to a PM/PEO or an operational unit for production, fielding, operations, maintenance, and/or support activities
Valley of death	A gap in development, production, or fielding when a technology has been demonstrated at technology readiness level 5 or higher and is technically ready to be incorporated into a system design or program and transitioned from development to production, fielding, operations, maintenance, and/or support activities
	Includes the period of time in which a business is seeking to transition a prototype or commercially available product to a DoD contract[b]

[a] See the organizational websites of DARPA, the Defense Innovation Unit, and AFWERX as of March 30, 2022.

[b] James M. Landreth, "Through DoD's Valley of Death," Defense Acquisition University, February 1, 2022.

Approach

We used a mixed-method qualitative approach to accomplish five tasks.

Task 1. Understand How Defense Innovation Organizations Seek to Accelerate Development and Adoption of Commercial Technologies

To understand how DIOs seek to foster innovation and accelerate the identification, development, and adoption of commercial technologies developed by the private sector for use by the U.S. military, we conducted a scoping literature review to identify DoD-related organizations involved in some way with the CTP, including DIOs as well as traditional DoD and component requirements, acquisition, and budgeting organizations. We analyzed applicable law, policy, and other information provided by the organizations, including existing innovation ecosystem maps and assessments related to the respective missions, authorities, processes, activities, organizational structures, human capital (both leadership and workforce), collaborative relationships, and resources of these various organizations. We supplemented the literature review with more than 50 interviews with more than 70 individual interviewees representing 18 defense innovation organizations, seven other innovation ecosystem stakeholders, and seven commercial businesses.

Task 2. Develop a Commercial Technology Pipeline Model

We developed a model of the CTP for analytic purposes, seeking to describe an idealized vision of the CTP that identifies the key activities, functions, and processes required to move a technology from idea to fielding under the current set of organizations and requirements, acquisition, and budgeting processes.

Task 3. Examine Whether and How the Defense Innovation Ecosystem Is Effectively Accelerating the Identification, Development, and Adoption of Commercial Technologies

We selected 12 DIOs to study in more detail, focusing on organizations that represented different missions, organizational perspectives, parent entities, and tenure and developed an organizational "deep dive" profile for each of the 12 DIOs that explains the CTP functions each performs and how it performs those functions. We conducted six case studies of new-entrant technology companies that have not previously sold to DoD in order to gain greater insight into the perspectives of these companies. Drawing on our analysis of data from all sources, we identified challenges, bottlenecks, gaps, and shortfalls that may be impeding the CTP and opportunities to strengthen these processes and approaches.

Task 4. Develop Alternative Approaches to Strengthen the DoD Commercial Technology Pipeline for Innovative Technologies

Using as baselines the CTP model and challenges, bottlenecks, gaps, shortfalls, and opportunities identified in task 3, the research team identified and explored alternative approaches to address these issues and to strengthen the CTP. We examined potential changes on the "supply" side (policies, processes, resources, and approaches that DIOs and commercial innovators use to supply innovative technologies to the military) as well as to the "demand" side (policies, processes, resources, and approaches that OSD and the services use to identify needs, develop requirements, and identify, develop, acquire, and integrate these technologies into military systems).

Task 5. Evaluate Alternative Approaches and Make Recommendations to Strengthen the DoD Commercial Technology Pipeline for Innovative Technologies

We developed and implemented an acquisition policy game that is based on a simplified model of the CTP. The game findings allowed us to understand some of the costs and benefits of specific reforms, which informed our recommendations.

Modeling the CTP

We developed a model of the CTP that identifies the key activities, functions, and processes required to move a technology from idea to fielding under the current set of organizations and requirements, acquisition, and budgeting processes (see Figure S.1). The CTP model can be used to understand key functions and activities, to standardize terminology to improve communication, and to allow stakeholders to see their roles and contributions to the CTP in context, as well as to inform stakeholders about the roles and contributions of other organizations. We used this idealized model as an analytic framework to identify stakeholder organizations, the CTP functions they perform, and their relationships to each other and the process.

Our CTP model is divided into three phases roughly corresponding to the maturity of a technology as it moves through the pipeline. Each phase contains a set of activities or functions that normally occur within that phase; these are the **core functions** of the CTP. Many of these core functions can be further subdivided into specific sets of activities, particularly those in the capability development phase, which includes science and technology (S&T) activities and research, development, test, and evaluation (RDT&E) activities. There is also a set of **enabling functions** that occur within and across each phase. These functions are defined in Chapter 3 (Table 3.1).

FIGURE S.1

Representation of the Commercial Technology Pipeline

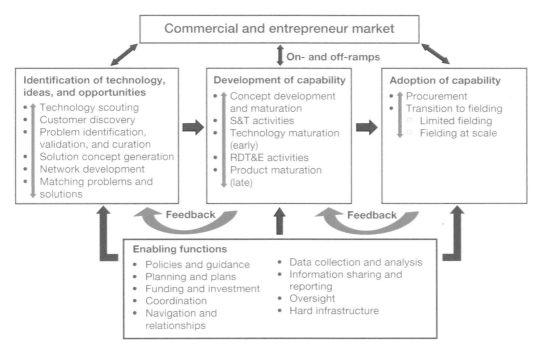

Key Findings

Understanding the CTP and the Role of DIOs

We examined the CTP as it exists today—the activities, functions, and processes by which the department currently identifies, develops, and adopts commercial technologies for military use, including innovation organizations' role in the CTP.

There is no single "pipeline" or pathway of technologies through the CTP. The CTP has many potential paths—combinations of activities within and between phases—from concept to fielding, and they are not necessarily linear or sequential. There are many on- and off-ramps at every phase, and feedback loops both within and between CTP phases. The path an innovative technology, product, or service takes from idea to fielding can differ depending on characteristics of the technology and business, financial considerations, and alignment with other DoD processes.

CTP functions are distributed across multiple stakeholders; no single organization performs them all. Given this, collaboration and handoffs between CTP stakeholders are essential to accelerate the identification, development, and adoption of commercial technology for military use.

DIOs perform multiple CTP functions but have limited ability to facilitate adoption. Many DIOs concentrate on early CTP functions (the identification phase), such as conducting acceleration programs and offering support to develop solution concepts, and they prioritize commercial technologies in defined technology readiness level (TRL) ranges. DIOs were established in large part to get new technology into the hands of the warfighter, but these organizations have limited ability to facilitate the adoption of new technology in the final stages: the actual procurement and fielding of technology. DIOs lack consistent buy-in from the traditional requirements and acquisition communities, in large part due to incentive structures that are misaligned between DIOs and these stakeholders.

Characteristics of a Well-Functioning CTP

Our review of CTP activities, functions, and processes led us to identify several characteristics of a well-functioning CTP:

- Stakeholders have a shared mission of pushing promising technology through the innovation life cycle from idea to fielding.
- Goals, objectives, and desired outcomes for the pipeline are established, understood, and shared by all CTP stakeholders.
- Each organization understands its roles and responsibilities and has an awareness of the roles and responsibilities of others.
- All key functions are performed by stakeholders (i.e., there are no gaps in individual functions), and handoffs of technologies from one function or activity to the next occur appropriately.
- Information sharing (promising technologies, available resources and programs, priorities/focus areas, collaboration opportunities) occurs appropriately across the CTP. Coordination and collaboration between CTP stakeholders and feedback mechanisms between CTP functions and stakeholders are established and functioning.
- Incentive structures for CTP stakeholders are aligned to CTP goals, objectives, and outcomes, including metrics and accountability mechanisms to check progress.

Challenges and Gaps in CTP Characteristics and Functions

Our research revealed a number of challenges in CTP characteristics and CTP functions. As detailed in Table S.2, the desired characteristics of a well-functioning CTP are not present: CTP stakeholders are not aligned to a shared mission or common goals, objectives, and outcomes. CTP roles and responsibilities are unclear, and there are significant gaps in key functions. There are no formal mechanisms or requirements for information sharing, coordination, or collaboration across CTP stakeholders. Finally, incentive structures for CTP stakeholders are not aligned to CTP goals, objectives, and outcomes, and there

TABLE S.2

Challenges and Gaps in CTP Characteristics

Desired CTP Characteristics	Challenge/Gap	Impact
Shared mission	CTP stakeholders are not aligned to a shared mission	Impedes technology transition and contributes to the "valley of death" problem
Common goals, objectives, outcomes	DoD has not established goals, objectives, or desired outcomes to guide CTP stakeholders	Stakeholder priorities and activities are not aligned, and collaboration is not incentivized, leading to inefficiencies and unnecessary costs
Clearly defined roles and responsibilities	Stakeholder roles and responsibilities are not well defined or understood	Creates confusion about who does what; leads to overlap of innovation organization activities and programming; leaves significant gaps in development activities and adoption functions
No gaps in key functions	*Gaps in key functions detailed in Table S.3*	*Impact of these gaps detailed in Table S.3*
Regular information sharing, coordination, and collaboration	Information sharing, coordination, collaboration, and feedback mechanisms between CTP stakeholders and functions are not established or functioning well	DoD does not realize value of problem curation, technology scouting, and other investments; CTP stakeholders do not know what others are focusing on, which impedes collaboration and contributes to the valley of death problem
Incentive structure aligned to CTP goals, objectives, and outcomes	Incentive structures for innovation organizations are not aligned to CTP goals, objectives, and outcomes	Key stakeholders lack incentives to collaborate and facilitate adoption of commercial technology; DIO-led problem curation, technology scouting, and early technology development are divorced from the traditional requirements and acquisition processes, contributing to valley of death problem
Incentive structure aligned to CTP goals, objectives, and outcomes	No DoD-wide metrics and accountability mechanisms to check progress against common goals, objectives, and outcomes	DoD unable to monitor and assess progress against goals; unclear how to evaluate return on investment

are no DoD-wide metrics and accountability mechanisms to check progress against goals. Stakeholders throughout the enterprise set their own priorities and objectives, which for the most part are not integrated or aligned.

We also identified a number of challenges and gaps in CTP core and enabling functions, as detailed in Table S.3. Identification of commercial businesses is unsystematic, and there is no institutionalized approach or enterprise solution to store and share data. DoD end users do not effectively communicate capability needs, and once identified, problems are not shared across CTP stakeholders. New entrants struggle to find clear points of entry into the defense marketplace, and once those opportunities are identified, most are difficult and costly to pursue. Businesses find DoD difficult to navigate, and once a business has made it

TABLE S.3

Challenges and Gaps in CTP Functions

Challenge/Gap	Impact	CTP Function
Identification of commercial businesses and technologies appears unsystematic	DoD may be overlooking promising technologies	Technology scouting
There is no institutionalized approach to outreach or enterprise solution to store and share data	Contributes to duplication of effort, inefficiencies, and missed opportunities; DoD not capturing the full value of labor-intensive outreach and networking with commercial businesses and DoD partners	Network development
DoD end users do not effectively communicate capability needs	Contributes to duplication of effort and inefficiencies (as users seek to solve related problems independently) and missed opportunities to target novel solutions to problems	Problem identification, validation, and curation
Problems are not shared across CTP stakeholders	Contributes to duplication of effort, inefficiencies, and missed opportunities to target novel solutions to problems; DoD not capturing the full value of labor-intensive problem curation	Problem identification, validation, and curation; solution concept generation; customer discovery; matching problems and solutions
Businesses cannot easily locate an initial point of entry into DoD	Industry unable to identify and access DoD opportunities and understand how their technology applies to DoD missions	Technology scouting, customer discovery, solution concept generation
Businesses find it challenging to understand and "speak" DoD	Industry does not understand what DoD needs or how their technology may be relevant	Customer discovery, solution concept generation, and (indirectly) development and adoption functions[a]
Businesses cannot easily identify the right DoD officials to provide feedback on their technology	Businesses lack feedback that is required to refine a technology's application within a mission space and develop and mature promising technologies	Customer discovery, solution concept generation, matching problems and solutions, and all development functions
There are no warm handoffs, clear routing, or obvious next steps for businesses	Businesses struggle to understand what is next and to identify subsequent funding and support; technology development and adoption may stall	All development and adoption functions
Suitable funding for continued development of technology is limited, and available funding is difficult to locate	Leads to gaps in funding and delays; technology development may stall; business may struggle to remain viable	All development and adoption functions
Most existing opportunities are overly burdensome to pursue	Burdensome processes are disproportionately costly for new entrants, deterring them from working with DoD	All development and adoption functions

Table S.3—Continued

Challenge/Gap	Impact	CTP Function
Many factors (including funding, misalignment with budget process, technical failure, and lack of a formal requirement or customer who is ready) contribute to the "valley of death"	Efforts to address the valley of death may fail because they focus on only one of these factors	Transition to fielding
The valley of death persists in part because no DoD organization has visibility into all CTP activities or responsibility for CTP outcomes; and gaps between various stakeholders hinder adoption	Technologies stall in the valley of death without a clear path to continued development or transition	Transition to fielding

[a] This gap ties to multiple development functions, including further development and maturation of solution concept, S&T activities, early-stage technology maturation, RDT&E activities, and late-stage product maturation.

through a particular program or organization, there are no warm handoffs, clear routing, or obvious next steps for businesses.

While DoD has funding available to support technology development, much of that funding is concentrated in the early stages of development, and there is limited support for testing and the proof-of-concept demonstrations that can help sustain a company. Perhaps most significant, we found that the valley of death persists because (1) no DoD organization has visibility into CTP activities enterprisewide or responsibility for CTP outcomes; and (2) gaps between innovation organizations, requirements and capability developers, end users, program managers and program executive officers, and procurement and decision authorities hinder adoption, and there are no incentives in place to encourage stakeholders to overcome these gaps. As a result of these and other challenges, technologies stall in the valley of death without a clear path to continued development or transition. While there may be legitimate reasons that a potentially useful technology becomes stalled in the valley of death (i.e., there are higher priorities for investment in a constrained budget environment), that outcome should be an explicit choice by informed decisionmakers, not a result of misaligned processes or incentives.

Recommendations to Strengthen the CTP

The research team developed and evaluated alternative approaches to mitigate identified challenges and strengthen the DoD CTP, drawing on data and analysis of the literature review, interviews, and the policy game. We assess that CTP throughput and effectiveness can be enhanced by policy levers that cultivate desired CTP characteristics, encouraging and incentivizing coordination and collaboration and striking a balance between organizational independence, free-market competition, and more centralized direction of the CTP.

Cultivate Desired CTP Characteristics

Foster a Shared Sense of Mission Across All CTP Stakeholders

We recommend that DoD identify ways to align CTP stakeholders—including those in the traditional acquisition, requirements, budget, and end-user communities—to a common understanding of the CTP and a shared mission of identifying, developing, and pushing promising technologies all the way through the pipeline. By fostering a shared mission, DoD can lay the foundation for the information sharing, coordination, and collaboration that is required for the CTP to function.

Develop and Promulgate Strategy, Plans, Policies, and Guidance for the CTP to Establish Common Goals, Objectives, and Outcomes While Supporting Flexibility in Execution

We recommend that DoD develop and promulgate DoD-wide strategy, policy, plans, and guidance for the CTP that set common goals, objectives, and outcomes, establishing common reference points and understandings of key concepts such as transition. Relatedly, we recommend that DoD couple this guidance with appropriate oversight and other incentives to encourage compliance. At the same time, DoD should continue to support flexibility and agility in the execution of department-level strategy, policy, plans, and guidance. In doing so, DoD can incentivize innovation organizations and other CTP stakeholders to pursue problems and technologies that are deemed to be of the highest priority by leadership and to share information, coordinate, and collaborate more readily.

Define and Communicate Roles and Responsibilities for the CTP

DoD should clarify and formalize roles and responsibilities for and relationships between CTP stakeholders, specifying which entities are expected to perform which CTP functions. Critically, DoD must ensure that traditional acquisition stakeholders understand their roles and responsibilities for CTP outcomes and that they view these roles as consistent with and a distinct part of their job. In doing so, DoD can lay the foundation for coordination and collaboration, minimize duplication and overlap, and ensure that all key functions are performed, minimizing gaps.

Facilitate Information Sharing, Coordination, and Collaboration Across CTP Stakeholders

DoD should establish mechanisms to improve information sharing, coordination, and collaboration across innovation organizations and other CTP stakeholders. For example, we recommend that CTP stakeholders share and institutionalize information relating to end-user problems and needs, new-entrant ventures, technologies and their potential applications, investment considerations and decisions, transition plans, and contacts both inside and outside DoD. If CTP stakeholders share information, coordinate, and collaborate, they can more efficiently move technologies through the pipeline to fielding.

Implement Incentive Structures, Including Metrics and Accountability Mechanisms, to Align CTP Stakeholders to CTP Goals, Objectives, and Desired Outcomes

DoD should realign incentive structures for all CTP stakeholders that are aligned to CTP outcomes. This effort should include developing meaningful metrics tied to the priority CTP outcomes that are designed to encourage collaboration and mitigate the factors that result in the valley of death phenomenon.

Strengthen CTP Identification, Development, and Adoption Functions

Develop More Rigorous, Comprehensive, and Coordinated Approaches to Technology Scouting

We recommend that DoD develop more rigorous, comprehensive, and coordinated approaches to technology scouting and share this information across the DoD enterprise. The information on technologies and businesses developed as a result of these activities should be widely shared across the DoD enterprise.

Encourage CTP Stakeholders to Share DoD Problems Across the Enterprise

We recommend that DoD encourage CTP stakeholders at all levels—requirements and capability developers, acquisition organizations (program managers, program executive officers, system commands), and especially end users at both the individual unit level as well as major and combatant commands—to share DoD problems across the enterprise. This will allow DoD to realize the value of problem curation, which is a labor-intensive, potentially valuable output from DIOs.

Establish a Comprehensive, Integrated, Searchable Portal for New Entrants to Access DoD Opportunities

DoD should establish a portal to serve as a common entry point for early-stage ventures that is comprehensive, easy to navigate, and fully searchable, building on the OSD innovation website.[1] Through this portal, DoD could explain how commercial technologies can be identified, developed, and adopted; explain different types of opportunities; and describe CTP stakeholder roles. The portal could also be a vehicle to collate and publicize funding opportunities, engage prospective users and program managers, and provide guidance on the types of information that companies should be prepared to share with the department.

Establish Navigation Support Services to Help New Entrants

DoD should establish navigation support services to help ventures with promising technology understand DoD, identify and apply for opportunities, conduct customer discovery, and understand end-user needs. In so doing, DoD may field more technologies.

---

[1] Office of the Under Secretary of Defense, Research and Engineering, "Business and Industry," website, undated a.

Assign Responsibility for Oversight of the CTP, Including Transition Outcomes

We recommend that DoD assign responsibility for overseeing the CTP and realizing CTP outcomes, including transition, to an organization that is provided with the authority and budget to develop and promulgate strategy, guidance, and plans; develop and implement metrics and accountability mechanisms; and facilitate information sharing, coordination, and collaboration across CTP stakeholders.

Provide Flexible Funding to Incentivize CTP Stakeholder Collaboration and Close Transition Gaps

We recommend that DoD establish a flexible funding pool to facilitate transitions and support the maturation of promising technologies that might otherwise stall in the valley of death, and then incorporate those technologies in a program of record where appropriate. Flexible funds should be able to be applied to any development or early production activity; they should not be tied to the normal planning, programming, budgeting, and execution cycle or to an approved requirement and controlled and allocated by the single organization assigned responsibility for the CTP. DoD will need congressional approval to implement this recommendation.

Implement DIO Best Practices

While DIOs cannot solve these problems on their own, DIOs should implement these best practices: (a) Be led by a DoD problem rather than by innovative dual-use technologies; (b) include transition planning as early as possible, even during accelerator or incubator programs, and revisit that transition plan continually; and (c) recognize that an engaged DoD customer is critical for success.

Considerations for Implementation

Our recommendations are designed to be interdependent and mutually supportive and should be implemented as a package. Importantly, these governance tools must be calibrated to preserve the agility of innovative businesses and DIOs and increase the agility of the traditional processes responsible for bringing new capabilities to the warfighter. DoD should carefully assess the costs, benefits, and unintended consequences of any potential change to the system.

Adopting innovative commercial technology for military use is not just a technical problem but also a cultural one. Innovative technology often requires adaptation—behavioral changes in how a problem is addressed. The willingness to make the changes that may be required to develop and use innovative commercial technologies is often a cultural issue and can constrain or hinder adoption of that technology. DoD should explore this aspect of technology adoption further to improve CTP outcomes.

Acknowledgments

This study would not have been possible without the assistance of many people within DoD and NDRI. First, we would like to thank our project sponsor, the National Security Innovation Network. In particular, the support and guidance of Abigail Desjardins, Greg Bernard, Jennifer Bird, Amy Schafer, and Morgan Plummer were critical to the project. We would like to thank representatives from the defense innovation ecosystem and early-stage ventures who spoke with us and participated in the acquisition policy game.

We would also like to thank several of our NDRI colleagues. We are grateful to the NDRI Acquisition and Technology Policy Center leadership team of Christopher Mouton, Joel Predd, Yun Kang, and Caitlin Lee for their guidance and support during the project. We are also grateful to three RAND Air Force Fellows—Lt Col (ret.) Frank Delsing, Lt Col Ryan "Ollie" Stallsworth, and Lt Col Ryan "Scar" Thulin—and RAND colleagues Bruce Held and Bill Shelton, whose feedback and insights were invaluable to this study. We thank Kristin Leuschner for her help summarizing our findings and improving the clarity of this report, as well as staff from Research Editorial and Production for their help guiding this report through publication.

Finally, we thank our reviewers, Timothy Bonds and Irv Blickstein, whose thoughtful reviews and comments on the research plan and final report undoubtedly enhanced the quality of this study. Any errors or omissions are the responsibility of the authors.

CHAPTER 1

Introduction

Technological superiority is a key element of the U.S. military's strength and advantage over its competitors. In order to maintain this advantage, the United States must promote and harness technological innovation, both within the federal government and in the private sector. This principle underpinned government-sponsored science and technology research during the Cold War and was a core element of the U.S. advantage during this period.

In more recent years, technology innovation has been driven by the commercial market. U.S. adversaries have access to that market, allowing them to reduce U.S. technological advantage. As China is aggressively modernizing and growing its military, the technological advantage the United States holds over its rivals has narrowed. To counter this challenge, the United States must continue to strengthen its national security innovation base by supporting defense-relevant research and development (R&D) and ensuring that these technologies and advances are adopted and fielded quickly to benefit the warfighter.[1] Department of Defense (DoD) direct investment in basic R&D remains critically important but is not sufficient to retain the U.S. technological advantage. Rather, DoD must find a way to identify, support, and leverage innovation relevant to national security problems that occur in the private sector, much of which is driven by the commercial marketplace and is conducted by individuals and entities that have traditionally not worked with DoD.

Recognizing this need, a number of defense innovation organizations (DIOs) have been created in the last two decades, both within the Office of the Secretary of Defense (OSD) and the military services, to bridge the gap between innovative private-sector organizations and the military, helping to foster communities of innovators and to accelerate the identification, development, and adoption of commercial technology, developed in the private sector, into the military. Many of the OSD and service-level organizations that make up the defense innovation "ecosystem," such as the Defense Innovation Unit (DIU), the National Security Innovation Network (NSIN), DEFENSEWERX, SOFWERX, the Air Force's AFWERX and Kessel Run, and the Army Applications Laboratory (AAL), were established independently of one another to address particular but often similar needs identified by their parent organization. This has led to a proliferated ecosystem of DoD innovation labs, hubs, and centers. Some observers have begun to ask how well these organizations have met their stated aims

---

[1] David Vergun, "Official Says DOD, with Help from Partners, on Cusp of Cutting-Edge Innovations," Defense.gov, November 8, 2021.

in promoting innovation and transitioning new technologies to the warfighter and whether the existing defense innovation ecosystem is optimally organized to support innovation and the transition of emerging commercial technologies for military use.[2]

Focus of This Report

DoD's National Security Innovation Network (NSIN) requested assistance from the RAND National Defense Research Institute (NDRI) to understand how defense innovation organizations can more effectively create and strengthen a "commercial technology pipeline" through which innovative commercial technologies can be identified, developed, and transitioned from the private sector to DoD for military use at both the joint and service level.

We sought to understand the following:

- What are the factors (e.g., functions, processes, relationships, coordinating structures) that support accelerated identification, development, and adoption of commercial technology?
- To what extent are those factors sufficiently present in the defense innovation ecosystem?
- How can challenges be mitigated so that any missing or insufficiently present factors can be generated and sustained?

To answer these questions, we sought to understand what defense innovation organizations are seeking to do, the types of inputs these organizations use, the activities they conduct, the outputs they generate, and they outcomes they are seeking. We then sought to identify the functions and other factors that are required to effectively accelerate the identification, development, and adoption of commercial technologies for military use. We identified challenges and gaps in current organizational structures and processes, and developed and tested alternative approaches to strengthen these structures and processes. Based on this analysis, we developed recommendations to strengthen the commercial technology pipeline. We used a mixed-method approach, described in Chapter 2, to gather and analyze data relating to these tasks.

Key Terms

"Innovation" is a frequently used term in the DoD, assumed to be desirable and rarely defined. The term is used to describe a wide range of products and practices. For the purposes of this project, we did not seek to define innovation in general terms; rather, we sought to understand what "defense innovation" means in the context of the defense innovation ecosystem.

---

[2] Vergun, 2021.

The team identified some themes and common elements in the way this phrase is used in the literature.[3] Definitions of defense innovation often, though not always, focus on a tangible product as an output. In this usage, "defense innovation" refers to the identification and development of new products, very often technologies, with potential military use. Many definitions emphasize the fielding of these products to users as new or improved capabilities. Relatedly, some definitions emphasize the "uptake" or "adoption" of such products. In this same vein, many definitions of defense innovation emphasize outcomes associated with the use of the new or improved product, such as increasing military effectiveness, shifting the balance of power, or satisfying an unmet military need. Other definitions emphasize the novelty of the innovation (e.g., how surprising or novel is it), distinguish between the different types of impact that the innovation might have (e.g., from sustaining and incremental to disruptive and even destructive), or emphasize the speed of innovation.

Each of these definitions excludes many types of innovation. We did not want to constrain or scope the study in this way. For purposes of this project, we have used an intentionally broad definition of "defense innovation," a phrase we use loosely to describe processes of generating and fielding technologies and other products, services, processes, or practices that are new or improved in the defense context.

Table 1.1 provides definitions of the key terms used in this report.

Structure of Report

The remainder of this report is organized as follows. Chapter 2 describes our approach. Chapter 3 explains the commercial technology pipeline (CTP) model and defines CTP activities, functions, and processes. Chapter 4 describes the gaps and challenges associated with current processes and approaches that we identified in this analysis. Chapter 5 describes and evaluates alternative approaches to strengthen the CTP, focusing on the alternatives we assessed through the acquisition policy game. Chapter 6 synthesizes our findings, recommends actions that DoD can take to strengthen the CTP, and discusses considerations for implementation.

---

[3] Phil Budden and Fiona Murray, "*Defense Innovation Report: Applying MIT's Innovation Ecosystem and Stakeholder Approach to Innovation in Defense on a Country-by-Country Basis,*" MIT Lab for Innovation Science and Policy, May 2019; Adam R. Grissom, Caitlin Lee, and Karl P. Mueller, *Innovation in the United States Air Force: Evidence from Six Cases*, Santa Monica, Calif.: RAND Corporation, RR-1207-AF, 2016; Jon Freeman, Tess Hellgren, Michele Mastroeni, Giacomo Persi Paoli, Kate Cox, and James Black, *Innovation Models: Enabling New Defence Solutions and Enhanced Benefits from Science and Technology*, Santa Monica, Calif.: RAND Corporation, RR-840-MOD, 2015; National Academies of Sciences, Engineering, and Medicine, *The Role of Experimentation Campaigns in the Air Force Innovation Life Cycle*, Washington, D.C.: The National Academies Press, 2016; National Academies of Sciences, Engineering, and Medicine, *Advancing Concepts and Models for Measuring Innovation: Proceedings of a Workshop*, Washington, D.C.: The National Academies Press, 2017; and U.S. Government Accountability Office, *Defense Science and Technology: Adopting Best Practices Can Improve Innovation Investments and Management*, Washington, D.C.: GPO, GAO-17-499, 2017.

TABLE 1.1

Definitions of Key Terms

Term	Definition
Defense innovation	The processes of generating and fielding technologies and other products, services, processes, or practices that are new or improved in the defense context
Defense innovation ecosystem	The set of defense innovation organizations, activities, functions, and processes that develop, produce, and field new or improved technologies and capabilities for military use
Defense innovation organization (DIO), innovation organization	DoD organizations that were founded to help the U.S. military pursue innovative ways to sustain and advance the capabilities of the force, including making faster use of emerging commercial technologies
	Includes long-standing organizations such as the Defense Advanced Research Projects Agency (DARPA) as well as the new generation of innovation organizations created in the last 15 years, such as DIU and AFWERX[a]
Commercial technologies	Technology originating from or matured by a commercial or private-sector entity
	Includes technologies or end products initially developed by the private sector (nondefense industry) without government funding and intended for the commercial marketplace
	Also includes technologies that were initially developed in DoD labs, licensed to the private sector to further develop and mature the technology or product through the commercial marketplace, and then sold back to DoD as a mature technology or product
Commercial technology pipeline (CTP)	The activities, functions, and processes by which DoD identifies, develops, and adopts commercial technologies for military use
CTP core function	Functions that occur within the three phases of the CTP: identification, development, and adoption
CTP enabling function	Supporting functions that occur both within and across the three CTP phases
CTP phase	One of three stages in the CTP: identification, development, and adoption
CTP stakeholder	DoD entities that have an interest, concern, or role in the CTP
	Includes requirements and capability developers, acquisition organizations (program managers, program executive officers, system commands), budget offices, and end users at both the individual unit level as well as major and combatant commands
New entrant	A commercial or private-sector entity that has not previously sold to DoD
	Includes but is not limited to "early-stage ventures," which are new entities (business, person, or group) with a new idea or concept for technology or an application of technology
Transition	A shift in responsibility or ownership of the technology product or system to a PM/PEO or an operational unit for production, fielding, operations, maintenance, and/or support activities
Valley of death	A gap in development, production, or fielding when a technology has been demonstrated at technology readiness level 5 or higher and is technically ready to be incorporated into a system design or program and transitioned from development to production, fielding, operations, maintenance, and/or support activities
	Includes the period of time in which a business is seeking to transition a prototype or commercially available product to a DoD contract[b]

[a] See the organizational websites of DARPA, the Defense Innovation Unit, and AFWERX as of March 30, 2022.

[b] James M. Landreth, "Through DoD's Valley of Death," Defense Acquisition University, February 1, 2022.

CHAPTER 2

Approach

The RAND NDRI research team used a mixed-method qualitative approach to understand how defense innovation organizations seek to accelerate the identification, development, and adoption of commercial technologies for military use; to understand and model the commercial technology pipeline; to identify gaps and challenges associated with current processes and approaches; and to identify and test alternative approaches that address those challenges.

Our research design included five tasks:

- Task 1. Understand how defense innovation organizations seek to accelerate development and adoption of commercial technologies.
- Task 2. Develop a commercial technology pipeline model.
- Task 3. Examine whether and how the defense innovation ecosystem is effectively accelerating the identification, development, and adoption of commercial technologies.
- Task 4. Develop alternative approaches to strengthen the DoD commercial technology pipeline for innovative technologies.
- Task 5. Evaluate alternative approaches and make recommendations to strengthen the DoD commercial technology pipeline for innovative technologies.

We will discuss each task in turn, along with the methods used to complete the task.

Research Tasks

Task 1. Understand How Defense Innovation Organizations Seek to Accelerate Development and Adoption of Commercial Technologies

We examined the defense innovation ecosystem to understand how these organizations seek to foster innovation and accelerate the identification, development, and adoption of commercial technologies developed by the private sector for use by the U.S. military. To complete this task, we conducted a focused open-source literature review, supplemented by interviews, to identify and analyze applicable law, policy, and other information provided by the organizations, including existing innovation ecosystem maps and assessments related to the respective missions, authorities, processes, activities, organizational structures, human capital (both leadership and workforce), collaborative relationships, and resources of these organizations.

Scoping Literature Review

We began with a scoping review of literature,[1] focusing on literature that described, analyzed, or critiqued the processes by which DoD identifies, develops, and adopts commercial technologies, defense innovation organizations, or the defense innovation ecosystem as a whole. Our goal was to understand

- how key terms such as "defense innovation," "technology," "commercial," and "dual-use" are defined and used by different stakeholders
- how DoD currently identifies, develops, and adopts commercial technologies for military use and which organizations are involved in these processes and activities
- what defense innovation organizations do and how they do it, which covers the inputs, activities, outputs, and outcomes of these organizations
- relationships, feedback mechanisms, and organizational contexts of defense innovation organizations.

The review was conducted by team members with expertise in national security, military acquisition and procurement, military technology, and structured analytic methods.

We identified relevant literature by conducting a search of a database of full-text articles in nearly 300 periodicals, academic journals, and other content pertaining to all branches of the military and the U.S. government. This search yielded more than 600 articles. After deduplicating, we reviewed the abstracts and full text of roughly 300 articles to identify relevant materials. We augmented this search by reviewing trade literature (e.g., *War on the Rocks*, *Defense News*), research and innovation organization websites—such as Defense Acquisition University (DAU), Naval Postgraduate School (NPS), and MITRE—and past RAND reports, and by querying RAND subject-matter experts who have worked in this space in the past. We used a "snowball" approach, working from initial sources to identify other relevant, publicly available sources.[2] We also obtained documentation directly from DIOs that describe their activities in more detail.

Through the literature review, we identified a myriad of DoD-related organizations that are involved in some way with the commercial technology pipeline, including DIOs as well

---

[1] Here, we define "literature" broadly to include research works from academic and other research institutions, industry- or sector-specific reports, and a variety of government documents. Scoping reviews can be appropriate in situations in which researchers are seeking to "identify the types of available evidence in a given field," to "clarify key concepts/definitions in the literature," to "identify key characteristics or factors related to a concept," or to "identify and analyse knowledge gaps" (see Zachary Munn, Micah D. J. Peters, Cindy Stern, Catalin Tufanaru, Alexa McArthur, and Edoardo Aromatris, "Systematic Review or Scoping Review? Guidance for Authors When Choosing Between a Systematic or Scoping Review Approach," *BMC Medical Research Methodology*, Vol. 18, No. 143, November 2018, art. 143.

[2] Julian Kirchherr and Katrina Charles, "Enhancing the Sample Diversity of Snowball Samples: Recommendations from a Research Project on Anti-Dam Movements in Southeast Asia," *PloS One*, Vol. 13, No. 8, August 22, 2018, e0201710.

as traditional OSD and component requirements, acquisition, and budgeting communities. Some documents we reviewed identified different sets of DIOs, depending on the specific focus of the work, but there seems to be a consensus that there are hundreds of such organizations within the innovation ecosystem.[3]

Given that a full examination of these organizations exceeded the scope of this research, we selected 12 DIOs for a "deep dive," as described under task 3 below.

Interviews

We conducted interviews with representatives of defense innovation organizations to understand how each organization seeks to identify and then accelerate the development and adoption of commercial technologies, how they implement their programs and initiatives, and the relationships between these organizations and other DoD entities responsible for acquisition and fielding of technology. In addition, we wanted to examine how these innovation organizations evaluate these programs and initiatives and explore the relationship between the approaches each organization uses for successful adoption of commercial technologies.

Interviewees were identified initially through literature and with input from the project sponsor. From there, to further identify potential organizations and interviewees, we used snowball sampling, a method in which each interviewee is asked to provide the name of at least one more potential organization or contact.[4]

The research team conducted interviews virtually between September 2021 and May 2022, seeking information on

- the background, scope, and mission of DIOs
- their activities
- how they measure outputs, outcomes, and success
- challenges to fulfilling their mission
- alternative policy or process change ideas that may address or mitigate those challenges.

The interviews with DIOs also supported an in-depth examination of how the DIOs performed their CTP-related functions and activities (described below).

All interviews were semistructured; that is, the research team guided the interview using broad questions related to the above topics, but interviewers encouraged interviewees to discuss relevant topics freely. This allowed interviewees to highlight issues and topics that interviewers were not aware of at the outset of the interview. Interviews were also conducted under assured confidentiality to encourage candid conversations.

---

[3] See, e.g., Golden Wiki, "Department of Defense Innovation Ecosystems," webpage, undated. The Innovation Steering Group of the Office of the Under Secretary of Defense, Research and Engineering (OUSD/R&E) constructed a website that lists many DIOs and related opportunities; see OUSD/R&E, undated a. See also Anne Laurent, "So Many Innovation Hubs, So Hard to Find Them," *GovExec.com*, September 11, 2019.

[4] Kirchherr and Charles, "Enhancing the Sample Diversity of Snowball Samples."

We conducted more than 50 interviews with more than 70 individual interviewees representing 18 defense innovation organizations, seven other innovation ecosystem stakeholders, and seven commercial businesses that had interacted with and participated in the activities of one or more DIOs (discussed under task 3 below). We conducted follow-up interviews with many interviewees.

Task 2. Develop a Commercial Technology Pipeline Model

We sought to understand the activities, functions, and processes, including relationships among stakeholders and coordination mechanisms, that are required to move commercial technologies over the innovation life cycle from concept to fielding for military use. Drawing on the prior task, we developed a notional CTP model that describes the core and enabling functions required to identify, develop, and integrate commercial technologies into DoD for military use. The process of adopting commercial technologies or products into defense systems includes the identification of candidate technologies, attracting nontraditional defense businesses to accept DoD funding or participate in DoD-sponsored demonstrations or programs, adapting technologies for military use, and transitioning technologies into existing or new defense programs for further development and fielding. We identified the characteristics of a well-functioning CTP. This model served as the analytic framework for subsequent tasks.

Task 3. Examine Whether and How the Defense Innovation Ecosystem Is Effectively Accelerating the Identification, Development, and Adoption of Commercial Technologies

The research team used the CTP model developed in task 2 to examine how a subset of defense innovation organizations are seeking to accelerate the identification, development, and adoption of commercial technologies developed in the private sector for use by the U.S. military. Drawing on the literature review and interviews described in task 1, we developed deep-dive profiles of 12 organizations and their missions, activities, functions, processes, and outputs in order to explain how they perform CTP functions. We identified other stakeholders whose participation and collaboration are required for technologies to move through the CTP. We also developed six case studies of nontraditional technology companies that interacted with one or more DIOs. Drawing on our analysis of data from all sources, we identified challenges, bottlenecks, gaps, and shortfalls that may be impeding the CTP and opportunities to strengthen these processes and approaches.

Defense Innovation Organization Deep Dives

To gain a better understanding of how DIOs carry out the functions required to facilitate discovery and access to commercial technology, drawing on the literature review and interviews, we developed a set of organizational deep-dive profiles of a subset of organizations that we selected for in-depth study. The profiles provide a richer picture of each DIO that we examined, describing their missions, the functions within the commercial technology pipeline they seek to carry out, and how they perform those functions.

We selected innovation organizations to profile based on the certain criteria. The organization had to perform a function that involves some aspect of the CTP, so an organization whose primary function is to perform innovation internally targeted at military applications (e.g., Air Force Big Safari or Lockheed Martin Skunk Works) would not qualify. We also wanted to ensure that we profiled organizations that served different types of customers, which could include a broad focus on defense-wide problems or service or domain-specific problems. We also chose not to focus on organizations that contribute to technology development and research but whose founding principles are not explicitly centered on defense innovation. Examples here would include organizations that run federally funded research and development centers (FFRDCs) or university-affiliated research centers (UARCs).[5] Using open-source research and consultations with the sponsor, and following leads uncovered through the literature review and interviews, we developed a list of more than 60 organizations that met the criteria above. The organizations that we identified are listed in Table 2.1.[6]

From this group of 72 organizations, we sought to identify a subset of organizations to examine in depth. While we were not aiming for a fully representative sample of organizations, we sought variation in the organizations identified to capture potentially unique features or functions. We wanted to include at least one organization from OSD/Joint Staff and the Departments of the Army, Air Force, and Navy and both new and more established organizations; furthermore, we selected organizations that had different areas of focus in the innovation life cycle (phases or steps from idea to fielding).

Based on these criteria, we selected the 12 organizations marked in bold and underlined in Table 2.1 for our in-depth analysis and deep-dive profiles. These are listed here for convenience:

- Army Applications Laboratory
- AFWERX
- ARCWERX
- Capability Prototyping/Rapid Reaction Technology Office
- DARPA (Defense Advanced Projects Research Agency)
- Defense Innovation Unit
- Doolittle Institute
- Joint Rapid Acquisition Cell
- National Security Innovation Network
- NavalX
- SOFWERX
- xTechSearch

---

[5] RAND operates four FFRDCs, including the one within which this research is taking place: Project Air Force, Arroyo Center, National Defense Research Institute, and Homeland Security Operations Analysis Center.

[6] We note that some of the organizations listed might not be considered by some observers to be part of the DoD innovation ecosystem, and it is likely that there are organizations that others would include that are not listed here.

TABLE 2.1

Candidate Organizations for Deep-Dive Profiles

DoD-wide	Dod-wide (cont.)	Dept of the Army	Dept of the Navy	Dept of the Air Force	Other
DARPA	Rapid Innovation Fund	Asymmetric Warfare Group	CNO Rapid Innovation Cell	AFVentures	Defense Entrepreneurs Forum
Defense Digital Service	**Capability Prototyping/ Rapid Reaction Technology Office**	**Army Applications Lab**	Office of Naval Research	**AFWERX**	In-Q-Tel
Defense Innovation Board	Small Business Innovation Research	75th Innovation Command	Marine Corps Rapid Capabilities Office	Air Force Rapid Capabilities Office	Applied Physics Laboratory
Defense Innovation Unit	Small Business Technology Transfer	Army Office of Small Business Programs	Marine Corps Warfighting Lab	Air Force Research Lab	Lux Capital
Defense Innovation Marketplace	Strategic Capabilities Office	Rapid Capabilities Office	Naval Postgraduate School	Air Force Techstars Accelerator	Maryland Innovation and Security Institute
DEFENSEWERX	TechLink	Army Research Lab (ARL)	Naval Research Lab	Allied Space Accelerator	National Center for Simulation
Federal Innovators Network/Salon	Trusted Capital Digital Marketplace	Basic Research Innovation and Collaboration Center	**NavalX**	**ARCWERX**	NATO ACT Innovation Hub
Hacking for Defense	Joint IED Defeat Organization (JIEDDO)	Army Venture Capital Initiative	Next Generation Logistics	**Doolittle Institute**	On Point Technologies
J-8 Innovation Cell	**Joint Rapid Acquisition Cell (JRAC)**	ERDCWERX	San Jose Innovation Unit (USMC)	Griffiss Institute	Second Front Systems
Irregular Warfare Technical Support Directorate	Manufacturing Innovation Institutes	Rapid Equipping Office		Kessel Run	Silicon Valley Defense Group
Joint Integrated Air and Defense Organization	**National Security Innovation Network (NSIN)**	XTechSearch		MGMWerx	SOCOM Acquisition Agility Office
Joint Artificial Intelligence Center (JAIC)	National Security Technology Accelerator			STRIKEWERX	**SOFWERX**
Joint Capability Technology Demonstration Office	OSD Innovation Steering Group			Wright Brothers Institute	Tech Grove

NOTE: Organizations in bold are examined in deep-dive profiles (Appendix A).

We conducted open-source research on each organization and contacted them by email to request one or more interviews with representatives of the organization. We conducted interviews with at least one representative, and in most cases multiple representatives, from each of these organizations. Each interview was conducted virtually through Microsoft Teams or teleconference for approximately 60 minutes. The interviews covered the topics mentioned earlier, including

- organizational background, mission, and focus (e.g., primary customer base)
- organizational activities and functions
- definitions and measures of success
- challenges faced and how they have been or can be addressed.

Once we had conducted interviews and completed open-source research, we developed a deep-dive profile for each DIO that provides basic information on the organization (e.g., date founded, mission, budget, customer base), the CTP functions it performs, how it performs those functions, definitions of success, and CTP-related outcomes. We provided each organization with a draft of the deep-dive profile for comment and to respond to additional questions. The profiles are included in Appendix A. Our analysis of the organizations and the CTP functions that they perform is discussed in Chapter 3.

Technology Case Studies

In our discussions with DIOs, we gained insight into how they viewed their roles in the larger defense innovation ecosystem and the functions they carry out. We wanted to also gain greater insight into the perspectives of nontraditional technology companies that are the primary focus of attention from the DIOs. By interviewing representatives of companies that have participated in DIO initiatives, we gained a fuller understanding of the challenges the companies faced, how they benefited from interactions with DIOs, and where there are opportunities for DoD to improve. Each company has unique experiences engaging with DoD, and their perspectives are not necessarily generalizable to the experience of others who have also engaged with DIOs. Nonetheless, the case studies reveal interesting insights as one source of data on the CTP and defense innovation ecosystem.

We were not aiming to identify a representative sample but to identify companies that had experience with different DIOs (and in some cases they had interacted with multiple DIOs). We identified potential case studies first by examining materials from the DIOs, such as annual reports and information on their websites. We also asked several of the DIOs that we profiled to recommend companies to contact, though we note that this approach naturally can introduce some potential bias in the sample because DIOs are more likely to recommend companies they believe had a positive experience in their interactions. In several cases, the DIOs themselves stated that some of the companies they recommended to us would have useful critiques and not just "good news" stories to convey. Although the selection criteria did not include factors such as maturity of the company or the type of technology area, the sample of companies varies along these dimensions also.

We conducted open-source research on each company to understand the technology they are developing, the customer base they are seeking to develop, and the company history. We requested not-for-attribution interviews via email with an explanation of the project objectives, the sponsor, and the topics we wished to cover in the interview. Interviews were conducted virtually through Microsoft Teams or teleconference for approximately 60 minutes, with at least one researcher and one notetaker participating. Companies often had multiple leaders and staff members participate in the interviews. We encouraged interviewees to provide their candid views on their experiences engaging with DoD and DIOs and found in speaking with representatives of these companies that they indeed had diverse experiences, not always positive, which indicate some of the challenges in supporting the CTP through the defense innovation ecosystem.

Table 2.2 lists the businesses that we profiled, a short description of their technology, and the DIOs with which they interacted. Findings from the cases are woven into Chapters 3 through 6, and the case studies are in Appendix B.

TABLE 2.2

Technology Case Studies

Business Name	Technology Description	DIO Interactions
Vita Inclinata	Hardware and software to stabilize loads lifted by cable in high winds	MD5,[a] AFWERX
Compound Eye	Software algorithm enabling humanlike vision using cameras	xTechSearch
Distributed Spectrum	Hardware-agnostic radio frequency (RF) spectrum monitoring system that can detect, classify, and localize RF signals	National Science Foundation (NSF) Small Business Innovation Research (SBIR), NSIN
Company X[b]	Small unmanned marine vessel and elevated mast	DARPA, U.S. Navy SBIR, RRTO
Black Cape	Platform-agnostic data analytics, machine learning, and artificial intelligence applications and services	SOFWERX
Xona Space Systems	Position, navigation, and timing (PNT) service using small satellites in low earth orbit (LEO)	DARPA, the National Geospatial-Intelligence Agency (NGA), U.S. Navy SBIR, NSIN, and the NPS

[a] MD5 was the former name of NSIN.

[b] This firm asked for anonymity.

Task 4. Develop Alternative Approaches to Strengthen the DoD Commercial Technology Pipeline for Innovative Technologies

Using the CTP model and the challenges, bottlenecks, gaps, shortfalls, and opportunities identified in task 3 as baselines, the research team explored alternative approaches to address these issues and to strengthen the CTP. We examined potential changes on the "supply" side (policies, processes, resources, and approaches that defense innovation organizations and commercial innovators use to supply innovative technologies to the military) as well as the "demand" side (policies, processes, resources, and approaches that OSD and the services use to identify needs, develop requirements, and acquire and integrate these technologies into military systems). The changes, or alternative approaches, that we considered had many options, such as the changes to the missions, authorities, processes, organizational structures, human capital (both leadership and workforce), collaborative relationships, and resources of existing defense innovation organizations; the creation or elimination of innovation organizations; and changes to the processes and structures through which defense innovation activities are overseen.

Task 5. Evaluate Alternative Approaches and Make Recommendations to Strengthen the DoD Commercial Technology Pipeline for Innovative Technologies

Finally, we developed and facilitated a structured analytic policy game to evaluate a subset of the approaches identified in the prior task, leveraging information collected and the analytic framework developed in prior tasks. Building on past RAND work using policy games to study how stakeholders might respond to the implementation of new acquisition policy,[7] we developed a bespoke game. By creating an artificial environment in which participants assigned to specific roles could make decisions and experience the consequences of their choices, the research team was able to observe some of the costs and benefits of prospective alternative approaches to inform our recommendations. The game was conducted virtually over two half days.

Based on the literature review, deep dives, and case studies, the team developed two notional policy regimes that incorporated alternative approaches to strengthen various aspects of the CTP that prior tasks had identified as challenges. We refer to each specific alternative approach—a change in some aspect of the CTP—as a policy lever. One set of policy levers represented a minimal change to the status quo, while the other represented a

---

[7] See Elizabeth M. Bartels, Jeffrey A Drezner, and Joel B. Predd, *Building a Broader Evidence Base for Defense Acquisition Policymaking*, Santa Monica, Calif.: RAND Corporation, RR-A202-1, 2020; and Joel B. Predd, John Schmid, Elizabeth M. Bartels, Jeffrey A. Drezner, Bradley Wilson, Anna Jean Wirth, and Liam McLane, *Acquiring a Mosaic Force: Issues, Options, and Trade-Offs*, Santa Monica, Calif.: RAND Corporation, RR-A458-3, 2021.

more significant set of reforms. The game design represented the major elements of our CTP framework (described in detail in Chapter 3). Players represented key CTP stakeholder organizations, including DIOs, program managers, and end users. One set of policy levers was played on each day. Stylized descriptions of 12 notional technologies were provided as input on each day. Players were given different numbers of investment tokens depending on their role in the game. On both days, the objective was to move technologies through the different activities that constitute the CTP. Players used their investment tokens to indicate interest in a technology and funding of a specific activity.

A more detailed description of the game design and game play is provided in Appendix C.

Based on an analysis of game results, supplemented by an analysis of the literature review and interviews, the research team developed recommendations to strengthen the DoD commercial technology pipeline.

CHAPTER 3

The Commercial Technology Pipeline and Defense Innovation Organizations

DoD has historically relied on the private sector for the development and production of weapon systems, materials, and other equipment the military needs.[1] Businesses with a history of contracting with DoD, as either prime contractors or subcontractors providing goods or services, are part of the defense industrial base. In the past, DoD and other federal government organizations such as the National Aeronautics and Space Administration (NASA) and the Department of Energy invested heavily in R&D activities, from basic research through system development and production. This investment largely drove the development and maturation of critical technologies and new capabilities incorporated into DoD's weapon systems and other equipment. More recently, private-sector investment in the commercial market has driven the development of key technologies. It is therefore increasingly important that DoD expand the industrial base from which militarily useful technology is drawn to include new entrants—commercial or private-sector entities that have not previously sold to DoD—to maintain technological advantage over adversaries. The new generation of DIOs was established to improve or accelerate the adoption of commercial technologies, defined as technologies originating from or matured by a commercial or private-sector entity.[2]

These new organizations join older defense organizations such as DARPA and military department labs that have always sought to attract new, innovative businesses and technologies to the defense sector. Many DoD initiatives over the last several decades have had the same objective, such as advanced or joint concept technology demonstrators (ACTD/JCTDs) or use of Other Transaction Authority (OTA) contractual instruments. Traditional DoD

---

[1] C. Todd Lopez, "Consolidation of Defense Industrial Base Poses Risks to National Security," Defense.gov, February 16, 2022.

[2] Note that this includes technologies or end products initially developed by the private sector (nondefense industry), without government funding, and intended for the commercial marketplace, as well as technologies that were initially developed in DoD labs, licensed to the private sector to further develop and mature the technology or product through the commercial marketplace, and then sold back to DoD as a mature technology or product.

acquisition organizations also include a focus on innovation, including component labs, warfare centers, and centers of excellence (CoEs), and acquisition programs generally. Many of these more mature and established defense organizations also seek to attract innovative nontraditional businesses and commercial technologies. While we include some of these organizations in this research, we have focused on their CTP-related functions.

All these organizations are part of what is often called the "defense innovation ecosystem" (DIE), a highly complex set of many government and private-sector organizations that interact in different ways to generate innovative technologies with potential military utility, understand the potential capabilities those technologies represent, and develop and field those technologies as a part of an acquisition program to realize those capabilities. The DIE consists of both the older and newer generation DIOs; the traditional requirements, capability development, science and technology (labs and CoEs), and acquisition organizations in the DoD components; innovation cells within operational units; and the commercial and defense industry, including both entrepreneurial/early-stage and well-established businesses. There is no official count of all these organizations, but they are thought to number in the hundreds. A subset of organizations in the DIE supports the CTP, which we have defined as the activities, functions, and processes by which DoD identifies, develops, and adopts commercial technologies for military use.

Developing a CTP Model

We developed a model of the CTP for analytic purposes. In developing this model, we have sought to describe an idealized vision of the CTP that identifies the key activities, functions, and processes required to move a technology from idea to fielding under the current set of organizations and requirements, acquisition, and budgeting processes. We use this model to identify stakeholder organizations, the CTP functions they perform, and their relationships to each other and the process.

Scope of the Model

As we discussed in Chapter 1, for purposes of the CTP, we have defined commercial technologies as those originating from or matured by a commercial or private-sector entity. A commercial or private-sector entity includes both new entrants to the DoD market (a business that has not previously sold products or services to DoD) and established defense contractors, as well as early-stage ventures (ESVs), which are new entities—a business, person, or group—with a new idea or concept for technology or an application of technology.

In this analysis, we focus on technologies from a subset of organizations:

- Technology originating from a new entrant (including but not limited to ESVs) that may have potential military utility. This includes dual-use technology or products with both commercial and military purposes, regardless of whether the technology or product originates in the commercial or defense sectors, and existing commercial off-the-shelf (COTS) products that are generally mature products in the commercial marketplace but new to DoD.

- Technology originating from a government lab (including DoD and component labs, warfare centers, and CoEs) and matured in the commercial marketplace. This includes technology or concepts that are initially developed by a government lab but then licensed or handed off to a commercial business for further development and maturation within the commercial market, with the intention of then selling the technology or product back to DoD.

While technically part of the CTP, we are not including technology developed by established defense contractors that are already part of the defense industrial base. Technology developed by a government lab that is never sold commercially is also excluded from this analysis. However, one output of the CTP is an expansion and increase in the viability of commercial businesses working with DoD. This could include an established defense contractor selling a somewhat altered military technology or product in the commercial marketplace. Though relatively uncommon, examples of this last case include the use of Global Positioning System (GPS) technology in many commercial products and services, as well as the sale of Hummer vehicles, which were originally military vehicles that were modified for and sold on the commercial market.[3]

COTS and dual-use technology are common terms in the acquisition world, with policies intended to encourage their use dating back decades. This includes relying on the commercial market to sustain the viability of a business that also sells products to the government. The concept of using the commercial market to refine and mature a technology or product originating within the DoD science and technology (S&T) community, perhaps including DoD investment in a commercial business, is a newer concept.

CTP stakeholders include organizations that focus solely on the commercial aspects of technology innovation as well as those parts of defense organizations that focus on commercial or dual-use technology with potential military application.

Overview of the CTP Model

Figure 3.1 provides a stylized overview of the CTP model developed for and used in this analysis.

Our CTP model is divided into three phases roughly corresponding to the maturity of a technology as it moves through the pipeline. Each phase contains a set of activities or func-

---

[3] Ben Barber, "Military's Hummer Shifts to Civilian Market," *The Christian Science Monitor*, January 24, 1994.

FIGURE 3.1

Representation of the Commercial Technology Pipeline

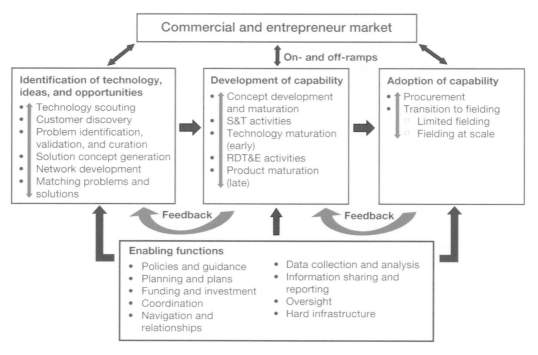

tions that normally occur within that phase (see Table 3.1 for definitions); these are the **core functions** of the CTP. Many of these core functions can be further subdivided into more detailed and specific sets of activities, particularly those in the capability development phase, which includes S&T activities and research, development, test, and evaluation (RDT&E) activities. There is also a set of **enabling functions** that occur within and across each phase. These enabling functions support the three different phases somewhat differently—that is, the policies governing each phase will be appropriate to the budget activity (BA) conducted in that phase or the applicable budget accounts (e.g., BA 6.1 to 6.7) that fund activities in which the technology or system is in a different level of maturity.

Our CTP model also includes interactions between the commercial and entrepreneurial market at any of the three phases that we define as the innovation life cycle (concept to fielding). The commercial marketplace is a source of technology or concepts, a mechanism for maturing technology, and a source of financial investment (e.g., selling a product in the commercial market or obtaining venture capital financing to support a business as it matures a technology or product).

The CTP process may look linear and sequential, and in some cases it is, as the term "pipeline" implies. And in general, the maturity of a technology or product increases sequentially through different activities over time. However, there are multiple feedback loops both within and between the three main phases. The CTP is not really a single "pipeline." There are multiple paths through which technologies are identified and matched to problems, and concepts and designs are matured, developed, tested, and adopted for military use. Technologies or products moving through the process do not necessarily go through every core activity; some may skip or repeat certain activities or may engage in multiple activities concurrently.

CTP Activities, Functions, and Processes

While there are many nuances in how a CTP function is performed, Table 3.1 captures the basic definition of each. DIOs and the traditional acquisition community—requirements developers, project managers (PMs), program executive officers (PEOs)—each perform some of the CTP functions that are necessary to facilitate adoption of commercially developed or dual-use technology by DoD, and we refer to each as a CTP stakeholder. No single CTP stakeholder organization performs all these functions themselves.

As mentioned earlier, enabling functions support each of the three phases in our CTP model, though the specific way each is implemented may be different by phase or core function. Enabling functions are truly enabling—core CTP functions cannot occur without funding (resources) and infrastructure (office space, information technology systems, test facilities) and an execution environment defined by policy and guidance.

There are other factors that may affect the CTP, including perceived threats; the state of the economy; and political support for businesses, technologies, or programs. Soft infrastructure,[4] including the financial and legal system and the climate or environment for innovation (e.g., geographic hubs such as Austin, Texas, Silicon Valley, and Cambridge, Massachusetts, with high concentrations of innovative new-entrant organizations and investors), provides the context within which the CTP operates and may define or constrain how a function or activity can be implemented. Finally, case studies of many successful technology innovations consistently point to the importance of the characteristics of key players—their drive, persistence, creativity, risk tolerance, subject-matter expertise, knowledge of the processes, and management expertise.[5]

---

[4] James Andrew Lewis, *Mapping the National Security Industrial Base: Policy Shaping Issues*, Washington, D.C.: Center for Strategic & International Studies, May 2021.

[5] The characteristics of key players is a factor in all of the case studies of commercial technology conducted as part of this research. This factor is significant in many weapon system cases studies as well. See, for example, Peter Garrison, "Head Skunk," *Air & Space*, March 2010; Lockheed Martin, "Missions Impossible: The Skunk Works® Story," webpage, undated.

TABLE 3.1
Definitions of CTP Functions

Core functions

Identification of technology, ideas, and opportunities

Technology scouting	Searching for new technologies and ideas and mapping technology trends; can include a review of or search for other DIO or industry data sets (e.g., SBIR contract awards database)
Customer discovery	Focused exploration and identification of potential DoD and/or commercial customers for a specific technology or product
Problem identification, validation, and curation	Identification, validation, and refinement of DoD problems for which innovative and nontraditional solutions may be appropriate
Solution concept generation	Generation or support to develop potential technology ideas or conceptual solutions to DoD customer problems
Network development	Focused development of specific digital and interpersonal networks as a result of outreach and communication; includes data capture in a customer relationship management (CRM)-type database
Matching problems to potential solutions	Aligning a curated DoD customer problem with one or more potential technology solutions

Development of capabilities

Further development and maturation of solution concept	Refining and maturing a technology solution concept and its application in a specific product
S&T activities (includes experimentation)	Conducting S&T activities associated with basic research, applied research, and advanced development of technology; includes prototype demonstration and experimentation to validate technology and solution feasibility; consistent with budget activity (BA) 6.1 to 6.3 funding
Technology maturation (early stage)	Initial development of early-stage technology, moving from concept to prototype; can include SBIR phase I to phase II contracting
RDT&E activities (including prototyping, user testing, and feedback)	Conducting late-stage development activities, including engineering and manufacturing development, system integration, and test and evaluation; includes more advanced (product-representative) prototyping and testing to obtain user feedback; consistent with BA 6.4 funding and above
Product maturation (late stage)	Refinement of technology into product application

Adoption of capabilities

Procurement	Production and purchase of a technology or system intended for fielding
Transition to fielding[a]	A shift in responsibility or ownership of the technology product or system to PM/PEO or operational unit for production, fielding, operations, maintenance, and/or support activities
Limited fielding	Small quantity of systems procured and fielded to single unit or few units; includes fielding residual capability from prototyping activities
Fielding at scale	Intended to be deployed across relevant service units; incorporation in a program of record (POR)

Table 3.1—Continued

Enabling functions

Policies, guidance	Development and implementation of statutory and regulatory policy and guidance specifying roles, responsibilities, and authorities for CTP functions
Planning and plans	Development of strategy and plans relating to the CTP; may include specification of goals, objectives, and milestones for CTP as a whole or for specific CTP functions
Funding and investment	Provision of funds to enable CTP activities
Coordination	Collaboration and partnerships among stakeholders in the CTP, including DIOs, accelerators, DoD customers, commercial venture capital (VC)
Navigation and relationships	Facilitating early-stage venture/business entry to DoD and interactions with DoD officials and potential customers; includes networking, mentoring
Data collection and analysis	Generation, collection, processing, storage, and use (analysis) of information about CTP operation to inform planning and decisionmaking
Information sharing and reporting	Relating to the CTP, includes networks of contacts (businesses, accelerators, mentors, DIOs, VC, DoD customers), data on problems, and data on potential technology solutions; also includes analysis measuring progress against strategic and tactical goals
Oversight	Provision of guidance and feedback to, and oversight of, defense innovation organizations
Hard infrastructure to support development	Provision of resources (funding, subject-matter experts), test range/ equipment, and other materials necessary to demonstrate capability

Other factors that can affect the CTP

External environment

Threat	Degree of perceived external threat as reflected in competition, crisis, and conflict
Market[b]	U.S. commercial market health, growth, technology trends, VC investment trends
Economy	Overall health of the U.S. economy, growth rates, inflation
Politics	Political environment, including industrial policy environment and government investment in commercial market

Other factors

Soft infrastructure	Aspects of society that facilitate and encourage entrepreneurship and innovation; includes legal, financial system, intellectual property rules, academic environment, entrepreneur culture
Early-stage venture and DoD customer enthusiasm, drive, persistence, critical thinking	Characteristics of the business or DoD customer that enable creative and critical thinking, intelligent risk-taking, persistence and consistency, problem solving (both process and technology)

[a] For interviews and in drafting the early version of this report, we used a slightly different definition of transition; namely, a "shift in responsibility or ownership of the technology product or system to PM/PEO or operational unit for fielding." We have since refined and clarified this definition.

[b] "Venture capital investment trends" was added after interviews were completed. This does not affect our analysis.

Utility of the CTP Model as a Framework

The CTP model provides a useful framework for understanding the many functions, activities, and relationships necessary to accelerate the identification, development, and adoption of commercial technologies. It identifies key functions and activities, standardizes terminology to improve communication, and allows stakeholders to see their roles and contributions to the CTP in context, as well as informing stakeholders as to the roles and contributions of other organizations.

The CTP model highlights some important observations that may otherwise be overlooked or not fully understood. For instance, the model makes clear that because no single DoD organization performs all the key functions and activities, collaboration, coordination, and information sharing among stakeholders is required to move a technology from concept to fielding. The model also suggests that feedback is important: DIOs need information and feedback from other stakeholders to determine whether they are generating useful outputs and ultimately contributing to CTP outcomes. Viewing DIOs through this framework can improve the understanding of what they do and how they do it and the critical roles of other CTP stakeholders such as requirements developers and PMs.

We apply this framework in the analysis that follows. The framework is also an output of this research and can be used to understand and analyze organizations, activities, functions, processes, policies, and resources, independent of our recommendations.

Characteristics of a Well-Functioning CTP

We applied the model to analyze the 12 innovation organizations that we profiled and to identify challenges and alternatives to strengthen the CTP. Based on this analysis, we identified characteristics of a well-functioning CTP, as shown in Table 3.2.

Taken together, these characteristics create the basic conditions for identifying, developing, and adopting commercial technology for military use. If these characteristics exist and are implemented efficiently and effectively, then the opportunities for accelerating CTP processes and increasing throughput (the number and rate of commercial technologies moving through the process) is improved.

DIOs and the CTP

Many organizations play roles in the CTP, a point that Deputy Secretary of Defense Kathleen Hicks emphasized on a trip to Silicon Valley in April 2022, noting that "the multiplicity of handoffs in our [acquisition] system that are required internally to get something from point A to point B is really striking."[6] She then noted that there was not a single-point solution to

---

[6] Valeria Insinna, "After Hearing Silicon Valley Complaints, Hicks Says No 'Magical' Fix to Acquisition," *Breaking Defense*, April 11, 2022.

TABLE 3.2

Characteristics of a Well-Functioning CTP

Characteristic	Description
Shared mission	CTP stakeholders have a shared mission of seeking to move potentially useful technology through the pipeline (innovation life cycle) from idea to fielding.
Common goals, objectives, outcomes	Goals, objectives, and desired outcomes for the pipeline are established to guide CTP stakeholders. This is not to say that the operation of the CTP should be centrally directed; rather, the goals of the overall system should be understood and shared by all stakeholders.
Clearly defined roles and responsibilities	CTP stakeholder roles and responsibilities are defined and understood. This means that each organization understands its roles and responsibilities and has an awareness of the roles and responsibilities of others.
No gaps in key functions	All key functions are performed by CTP stakeholders (e.g., no gaps in individual functions), and handoffs of technologies from one function to the next occur appropriately. Not every function is performed by a single entity, but all CTP functions appropriate to a particular technology must be performed by some organization.
Regular information sharing, coordination, and collaboration	Information sharing (relating to promising technologies, available resources and programs, priorities and focus areas, and collaboration opportunities) occurs appropriately across the CTP, improving the efficiency of the CTP and generating more opportunities for potentially useful technologies or products to be identified and fielded. Coordination and collaboration between CTP stakeholders are occurring, and feedback mechanisms between CTP functions and stakeholders are established and functioning.
Incentive structure aligned to CTP goals, objectives, and outcomes	Incentive structures for CTP stakeholders are aligned to CTP goals, objectives, and outcomes, including establishment of metrics and accountability mechanisms to check progress against goals. At a minimum, this means that the incentives in place for CTP stakeholders do not encourage decisions that conflict with CTP goals and objectives. Metrics for both overall CTP outcomes and for measuring the outcomes of specific activities at specific stakeholder organizations can be used for oversight and to ensure accountability.

addressing the challenge of acquiring innovative technology, echoing insights we provided in the previous section. DIOs are not the same; they carry out different functions in different ways, focus on different problems and customers, and have developed different core competencies as part of the CTP. There are some similar attributes across many DIOs, however, particularly those created in the past decade.

In our scanning of the defense innovation ecosystem, we identified many organizations, but only a subset of those is directly identifying, supporting, maturing, and transitioning promising commercial technology through the CTP. Although we could not analyze every DIO, as noted in Chapter 2, we sought to identify a sample of DIOs that is broadly representative of the types of organizations most clearly aligned with the CTP. Some organizations we did not examine may carry out CTP functions in different ways from those we examine here, therefore our findings on DIOs and the CTP may not be generalizable to the entire defense innovation ecosystem.

In Table 3.3, we map the functions of the CTP to the DIOs for which we developed profiles.[7] The CTP functions do not necessarily cover all activities that a DIO performs (i.e., activities associated with defense-specific technology) but include many of them.

---

[7] The DIO profiles, including much more detail on each organization and their CTP functions, appear in Appendix A.

TABLE 3.3

Mapping Innovation Organizations to CTP Functions

CTP Function	Defense Innovation Organization											
	AAL	AFWERX	ARCWERX	CP/RRTO	DARPA	DIU	Doolittle	JRAC	NavalX	NSIN	SOFWERX	xTechSearch
Identification of technology												
Tech scouting	X	X	X	X	X	X	X		X	X	X	X
Customer discovery		X		X		X	X			X		X
Problem identification	X		X	X	X	X		X	X	X	X	
Solution generation	X	X	X		X	X			X	X		
Networking	X	X	X			X	X		X	X	X	X
Matching	X	X	X	X	X	X				X		X
Development of capabilities												
Solution concept development and maturation	X	X	X			X		X		X		X
S&T activities		X		X	X						X	X
Technology maturation (early stage)	X	X	X		X	X	X		X	X		X
RDT&E activities	X		X	X		X					X	
Product maturation (late stage)		X			X	X			X			
Adoption of capabilities												
Limited fielding			X	X	X	X		X	X			
Fielding at scale	X			X	X	X						
Enabling functions												
Policy								X				
Planning												
Funding	X	X	X	X	X	X		X	X	X		X
Coordination	X	X	X		X	X	X	X	X		X	X
Navigation and relationships	X		X	X	X	X	X		X	X	X	X
Data and analysis	X											
Information sharing and reporting			X					X				
Oversight						X		X				
Hard infrastructure		X		X	X	X					X	

DIOs in our sample tend to perform early-stage core CTP functions, particularly in the identification phase. A review of the objectives and missions of the DIOs we examined suggests that the new generation of DIOs was established, in part, to perform the core functions (and some enabling functions) in the identification phase and as an alternate conduit into the DoD acquisition world.[8] Transitioning technologies further through the CTP requires the engagement of other stakeholders. Many of the CTP functions with limited DIO coverage—including important enabling functions—are owned principally by other stakeholders (e.g., PMs, requirements developers, end users, other OSD/service organizations).

Most DIOs Perform Multiple Functions to Identify Promising Technology

DoD has established DIOs because the traditional acquisition pathways are considered too complex, rigid, and burdensome to attract and identify early-stage ventures and companies that are not already known to DoD and the interesting technology they may be developing. All but one of the DIOs we investigated in-depth conduct some form of technology scouting.

Technology scouting can come in multiple forms: more traditional SBIR solicitations (a significant portion of AFWERX's approach), posting requests for proposals or information to SAM.gov, and newer approaches such as hosting challenges on hard problems and "colliders"[9] to bring innovators together to spark new ideas and approaches.[10] Some innovation organizations have established an organizational structure with regional hubs or collaboration spaces, such as NSIN (headquartered in Arlington, Virginia, with a regional network team and university program directors across the United States), DIU (offices in Silicon Valley, Boston, Austin, the Pentagon, and Chicago), and NavalX (Tech Bridges, offering a collaboration space in a commercial business space in locations across the United States and in London and Yokosuka, Japan.).[11] DIOs also conduct outreach by giving talks to local educational institutions, partnering with technology centers and schools, recruiting interns to conduct market research, and advertising activities and events on their websites, social media channels, and through email distribution lists. For example, DARPA hosts regional events

---

[8] See Appendix A for detail.

[9] MITRE defines a *challenge* as "a single or recurring contest or competition aimed at solving problems where emerging technologies have the potential to provide nontraditional solutions, or to expand the pool of participants to address critical issues." Challenges "may offer cash prizes or may be part of a broader challenge-based acquisition (ChBA) strategy that may result in a government contract." MITRE, "DoD Innovation Ecosystem," website, undated. *Colliders* are government-sponsored events, such as the Air Force's Spark Collider and the Defense Logistics Agency's R&D Industry Collider Day, designed to bring private-sector innovators and defense innovation organizations together to network, enable the DIOs to share new opportunities with industry, and allow innovators to present ideas to potential sponsors.

[10] Interview 26, Interviewee 33, September 24, 2021.

[11] Defense Innovation Unit, "Defense Innovation Unit (DIU): Who We Are/Our Mission," webpage, undated c; National Security Innovation Network, "Our Regions," webpage, undated d; NavalX, "Tech Bridges," webpage, undated d.

with R&D universities, seeking to connect DARPA leadership to scientists, innovators, and defense leaders.[12]

Discovering interesting technology also requires understanding the potential customers for that technology and the problems they face. Innovation organizations generally recognize that starting with technology and seeking to identify prospective customers for it—technology push—is difficult and unlikely to result in a successful transition.[13] As one innovation organization interviewee noted, "Matching [my organization's] ideas to the Army can be challenging because there is no single point of entry to the Army, and it can be difficult to gain traction with stakeholders if the idea is outside the Army's modernization priorities."[14] Another said, "We realized that we could do lots of cool widgets and prototypes. But then it goes to the prototype graveyard, which is a waste of taxpayer dollars. Our mission is to get it into the hands of warfighters. Without a clear DoD problem set [to start from], [our efforts to identify interesting technology] doesn't unlock scale at the end."[15]

Given this, many innovation organizations seek to be guided by high-priority end-user problems and devote considerable time and effort to identifying and curating problems from across the department. Innovation organizations often attempt to identify a prospective DoD customer or "transition partner," then seek to understand that customer's needs, support development of solution concepts to address those needs, and identify technologies that may meet those needs.

DIOs can develop relationships across their customer base through official channels or more informal networking across a military department, DoD component, or the DoD enterprise. The Army Application Lab is embedded in Army Futures Command, a four-star command with a mission of identifying and developing future capabilities for the Army. xTechSearch is chartered by the Assistant Secretary of the Army for Acquisition, Logistics, and Technology—ASA(ALT)—and therefore can create direct links to the program executive offices and program management offices throughout the Army. NavalX has established "Tech Bridges" that align to naval districts to ensure each Navy installation and command has a dedicated innovation hub. Other DIOs are deliberately established external to the existing organizational infrastructure to support a culture of innovation that is a step removed from the home institution's way of thinking about problems but which makes the linkages to end customers harder to establish and maintain. The Defense Innovation Unit established its first offices in Silicon Valley to be closer to an area of technology innovation and adopted the style, vocabulary, and techniques of private-sector innovators. DIU has since expanded to establish a presence in other technology hubs such as Austin, Boston, and Chi-

---

[12] Defense Advanced Research Projects Agency, "DARPA Forward," webpage, undated b.

[13] Interview 8, Interviewee 9, August 16, 2021; Interview 19, Interviewee 21, September 15, 2021; Interview 11, Interviewee 12, August 25, 2021.

[14] Interview 10, Interviewee 11, August 2, 2021.

[15] Interview 11, Interviewee 12, August 25, 2021.

cago. SOFWERX is under the DEFENSEWERX umbrella but maintains close links to special operations command (SOCOM) and the SOCOM program offices.

Some problem definitions come through a formal process, such as the Joint Urgent Operational Need submission process from combatant commands that informs the work of the Joint Rapid Acquisition Cell (JRAC), while others engage through looser affiliations and connections such as SOFWERX's engagement with SOCOM's 12 program executive offices and AAL's discussions with Army Futures Command's cross-functional teams (CFTs). The DIO can work with a DoD customer to craft their problem statement in a way that will resonate with an audience that is not typically steeped in military knowledge and culture. In many cases, the form of innovation must fit into one or more larger programs or initiatives where a small company is not likely to be competing for the overall program but could provide valuable features to enhance a program. NSIN, SOFWERX, ARCWERX, and NavalX are all examples of DIOs that conduct outreach to potential customers to learn about their problems and help shape them in a way that will translate to the commercial world, working with their customers in DoD to refine problem statements into a language that nontraditional performers will understand. The organizations try to understand the full spectrum of the problem and a potential customer's requirements. As one DIO put it, "We sit down with [the customer] on requirements, do the stakeholder mapping, [seeking to understand] their capability gaps, how will they use it. . . . [We want to] cure the disease, not just the symptoms."[16] AAL aligns its program managers to Army Futures Command's CFTs, an approach that is designed to create an ongoing dialogue between the PMs and the CFTs so that PMs understand Army modernization priorities and the challenges the CFTs are looking to address. Additionally, AAL engages operational units regularly to solicit problems that can be refined for solution solicitation.[17]

Solution concept generation occurs in many ways that engage the ultimate customer or end user to a greater or lesser extent. ARCWERX, for example, seeks to educate Air Force reservists and guardsmen to help drive innovation in their home units by teaching them techniques such as design thinking and by taking advantage of their civilian skills and experience to create innovative solutions.[18] OSD's CP/RRTO[19] shares promising technologies across a broad DoD stakeholder base and will match potential end users with the company to explore how to apply the technology or engage the potential customer in a government-only meeting to discuss how to potentially guide the technology to a prototype, whether and how the technology can be integrated into a program, and ways to cofinance with a program or end user.[20] DIU employs a two-step approach that starts with inviting companies to pitch

---

[16] Interview 21, Interviewee 25, September 16, 2021.

[17] Interview 17, Interviewee 19, September 24, 2021.

[18] ARCWERX, "Educate," webpage, undated.

[19] Until 2022, the office was known as the Rapid Reaction Technology Office (RRTO). It has been renamed Capability Prototyping (CP). We refer to the office as CP/RRTO throughout the report.

[20] CP/RRTO information to authors, April 27, 2022.

potential solutions based on defense problems the DIU has crafted with relevant DoD components. DIU (or DIU and prospective customers) select interested businesses to pitch their proposed solution to a defined problem. DIU selects a subset of these to move into a negotiation phase for a full proposal. DIU then uses OTA for prototyping to award an agreement to support prototyping of the solution. DIU notes that this process works best when solutions are well defined and mature enough to move to prototyping in one to two years.[21] NSIN's programs, *Propel* and *Starts*, support solution concept generation in different ways: Propel provides companies with training and access to DoD problem owners for customer discovery sessions to better refine their product offering to the DoD use case. Starts brings start-ups together for a pitch day to introduce the potential solutions to end users, with winning pitches receiving either a financial award through a prize authority or a contract to further develop their solution.[22] Both programs include venture capital and other private investors who may choose to also invest in the companies.[23]

Many DIOs match a specific problem identified by a DoD customer (i.e., an operational unit) with one or more potential solutions. CP/RRTO solicits short three-page white papers from innovators and convenes operators and technical experts from across DoD (including combatant commands, the services, and other DIOs) to review the proposals and identify potential military applications and customers. NavalX's Tech Bridges are aligned to naval districts to learn about challenges the Navy is facing and develop a regional innovation network of companies, research institutions, and the other military services.[24] DARPA's approach, on the other hand, is driven by program managers who pitch an area of research to the DARPA leadership and then hold open days and post broad area announcements online to solicit interest. DARPA programs are not typically designed with a specific end user or customer in mind but rely on service liaisons to help identify potential customers in their home institutions.[25]

All DIOs Support the Maturation of Technology, but Often Focus on Different Stages or Means

More than half the DIOs we examined in-depth provide support to mature solutions. NSIN's Propel program, for example, combines mentorship and potential access to follow-on funding to mature technologies in the middle tier of technology readiness levels (TRLs), TRL 4–6. The Army Application Lab uses multiple approaches to mature solutions, including

---

[21] Interview 11, Interviewee 12, August 25, 2021.

[22] NSIN information to authors, June 30, 2022.

[23] NSIN, "Accelerator Portfolio RNT Guide with Branding," February 15, 2021, on file with authors; "The Mission Model Canvas," Starts April 25, 2020, designed by Joey Clark (Starts MMC v4_0), on file with authors.

[24] NavalX, undated d.

[25] Defense Advanced Research Projects Agency, "Our Research," webpage, undated h; Interview 14, Interviewee 16, September 2, 2021.

in a single cohort, to bring multiple companies together to work on complex problems. Each of the AAL programs includes the potential for multiyear funding. The Army's xTech program runs challenges that integrate Army customers into the judging and feedback process so that companies benefit from better understanding the Army's needs and how their evolving technology and concepts need to mature to address those needs.

Some DIOs support prototyping. As one example, CP/RRTO has three primary funding lines that support development of prototypes,[26] including both proof-of-principle prototypes to "explore the art of the possible" and operational prototypes that are intended to move technologies across the valley of death and into a service program of record (POR).[27] These prototyping programs ranging from one to three years, and funding ranges from under $1 million to up to $15 million for operational prototypes.[28] Working on behalf of a government sponsor with specific needs, CP/RRTO also conducts targeted outreach to small business and nontraditional business partners,[29] soliciting proposals for projects to develop a "quick win for the joint warfighter" by prototyping a technology, usually a yearlong project funded at less than $1 million.

Many DIOs also facilitate proof-of-concept demonstrations based on a prototype. This is a critical step; it demonstrates both that the technology does what was intended and that the capability has military utility. NavalX tests prototypes with Navy personnel and proposed end users, but late-stage product maturation requires transitioning the validated prototypes to an acquisition organization. Most DIOs do not have the funding required to sponsor larger-scale live-fire exercises themselves; they can, however, advocate to include specific technologies or products in a live exercise run by an operational component. ARCWERX's funding is for operations and maintenance (O&M), which restricts its ability to fund any RDT&E efforts including prototyping. CP/RRTO is an exception. CP/RRTO funds three demonstration venues and makes them available to companies to demonstrate promising technologies.[30] The results from these experiments and demonstrations are made available to the government.[31] In fiscal year (FY) 2021, CP/RRTO supported demonstrations of 219 technologies to a total of 715 DoD/government attendees.[32] NSIN, since the inception of the Adaptive Threat Force (ATF) program in 2018, has engaged over 15,000 service members from the Marines,

---

[26] Rapid Reaction Technology Office (RRTO)/Capability Prototypes (CP), document, January 2022, on file with authors (hereafter RRTO/CP, January 2022). A fourth funding line was not in the fiscal year (FY) 2021 budget.

[27] RRTO/CP, January 2022; Rapid Reaction Technology Office, "Overview," slide presentation, February 2021.

[28] RRTO/CP, January 2022; RRTO, 2021.

[29] RRTO, 2021.

[30] CP/RRTO information to authors, April 27, 2022; RRTO, 2021.

[31] CP/RRTO information to authors, April 27, 2022.

[32] Rapid Reaction Technology Office (RRTO)/Capability Prototypes (CP), January 2022.

National Guard, Army, and special operations forces (SOF) and field-tested over 250 technologies at live, force-on-force events.[33]

There are many technology R&D activities,[34] from S&T support through RDT&E and product maturation, and we have not tried to characterize all of the ways that DIOs perform these functions. All the DIOs put emphasis on early-stage concept and technology development, though the degree of emphasis varies. AFWERX support to S&T comes through early-stage research supported by SBIR contracts. AAL also uses SBIRs but can also use programmed RDT&E funding to support prototyping, technology development, and demonstrations up to TRL 6.[35] The Doolittle Institute, conversely, only facilitates connections for small businesses to SBIR phase I and II opportunities, but it has run challenges that award prizes for successfully demonstrating a prototype to address the challenge, such as the Metal Additive Manufacturing Challenge held in 2021.[36] NSIN is most focused on early-stage concept development and solution maturation, while NavalX tends toward the later stages of technology maturation (particularly RDT&E and product maturation). In seeking to distinguish activities performed by certain DIOs, we note that funding mechanisms may restrict use of monies for certain stages of technology development.

Few DIOs Play a Significant Role in Fielding of Technology Where Other Stakeholders Have More Influence and Means

DIOs were established in large part to get new technology into the hands of the warfighter, but they have limited ability to facilitate the adoption of new technology in the final stages: the actual procurement and fielding of technology, whether at scale (e.g., across a service) or in a more limited fashion (e.g., for an operational unit such as a Brigade Combat Team or Marine Expeditionary Unit).

A prototype technology that has demonstrated some utility can be given to an operating unit for use (fielding the prototype represents residual capability from the perspective of the prototyping activity), but this may not lead to the broader service or component choosing to acquire the technology at scale. Many companies can successfully compete for and participate in multiple DIO events and competitions and garner follow-on contracts to continue maturation of technologies. Companies like United Aircraft Technologies and LiquidPiston, for example, have been xTech finalists, won SBIRs from the services, and have (in the case of United Aircraft) concluded partnerships with more traditional defense contractors (Bell

---

[33] NSIN information to authors, June 30, 2022.

[34] Note that we are using research and development here in a general sense to cover the spectrum of technology development from early-stage basic research to stages when technology is ready for prototyping, testing, and evaluation. We use S&T and RDT&E to refer more specifically to those activities where appropriate.

[35] Army Applications Laboratory, "What We Do," webpage, undated e.

[36] Doolittle Institute, "Grand Challenge: Metal Additive Manufacturing," webpage, undated a.

Helicopters). Companies working with SOFWERX have demonstrated several technologies that were then consigned or contracted to government customers, including a lightweight vehicle that can operate as a medical gurney in rough terrain and algorithms to improve imagery analysis that is now used by the National Geospatial-Intelligence Agency (NGA). But in many cases, the adoption of the technology relies on finding a program or other customer to provide funding for adoption, whether through limited fielding or fielding at scale. Several interviewees noted this requires careful planning from the start, as detailed in Box 3.1.

BOX 3.1

Teeing Up Successful Transition

"Seventy percent [of successful transition] is getting it right from the beginning. Another 20 percent is keeping the programs short and moving quickly, so we deliver it before the technology gets stale or the requirements/needs change. Ten percent is looking for additional transition partners while the program is executing."

> Jon Lazar, director, Capability Prototyping/Rapid Reaction Technology Office, June 6, 2022.

Many organizations are seeking to "bridge the valley of death" and overcome transition-related challenges by identifying a "transition partner" early in the process. As one successful example of this approach, CP/RRTO seeks to "invest where others do not," filling "seams, cracks, and fissures," and to secure cofinancing from other stakeholders and select technologies with a "clear transfer/transition path."[37] Box 3.2 provides an example of this work. CP/RRTO builds partnerships with end users and programs at the outset of the prototyping process to facilitate later adoption and fielding. As they work with companies and government partners, CP/RRTO fosters and supports relationships and partnerships with those transition partners to increase adoption rates of technologies that CP/RRTO has supported. CP/RRTO also supports development of technologies where the government transition partner and company are already connected, including companies that have identified government sponsor-transition partners and government sponsors that have identified a company with promising technology and solved the programmatic issues but need additional financial support to develop a prototype. Together, those things set conditions for limited fielding and fielding at scale. CP/RRTO transitioned 60 percent of technologies from their prototyping to "capability delivery," meaning that the prototypes "were delivered to program of record or joint warfighter as a capability, or significant component of a capability."[38]

---

[37] RRTO, 2021.

[38] Rapid Reaction Technology Office (RRTO)/Capability Prototypes (CP), "Transition Summit Excerpt," undated, on file with authors

---

BOX 3.2
Bridging the Valley of Death

"RRTO provided $2.037 million in prototype funding to Maritime Applied Physics Corporation (MAPC) for their concept for automatically deploying and retrieving a parafoil system with their large jet ski platform, the Greenough Advanced Rescue Craft (GARC). This system allows for greater digital communications and sensor connectivity. RRTO, recognizing the potential for payoff, invested early to develop this capability and mature the idea so it could eventually be demonstrated for its feasibility. Because of RRTO's support, MAPC successfully transitioned to the Navy's program executive office and they identified this technology as a solution for a program of record. RRTO's funding was key to finishing RDT&E efforts, eventual transition, maturation, and production under the program of record."

CP/RRTO information to authors, April 27, 2022; see also Xavier Vavasseur, "U.S. Navy Successfully Tests GARC/TALONS for LCS MCM Mission Package," *NavalNews.com*, November 26, 2019.

Funding structures (e.g., color of money) are one cause for challenges in transitioning technology. Organizations like AFWERX that use SBIRs to support technology development do not have funding to support transition or the follow-on maturation of a technology once a SBIR phase I and II award is complete. SBIR phase III does not in itself come with any SBIR/Small Business Technology Transfer (STTR) funding (although it does allow for other sources of U.S. government funding), so it is incumbent on the company and the customer to determine how to transition the technology. DIU uses OTA for prototype development, which can lead to some transition of demonstrated technology, but OTA is limited in funding ($500 million) and is an alternative contracting mechanism; it is not intended to replace formal acquisition processes. A sole-source production OTA can be used if notice is provided with a prototype OTA solicitation.[39]

DIOs' Coverage of Enabling Functions Is Limited

The DIOs we examined perform few enabling functions. A number of DIOs perform navigation and coordination functions, which is one reason the new generation of DIOs was established: to help businesses with innovative technologies understand and gain entry to the DoD market. Many DIOs provide funding to support some further refinement of solution concepts, and some fund prototypes or proof-of-concept demonstrations, though most often

---

[39] Ellen M. Lord and Michael D. Griffin, "Definitions and Requirements for Other Transactions Under Title 10, United States Code, Section 2371 b," memorandum, November 20, 2018.

in relatively small amounts (challenge prize, SBIR phase I or II) and limited to early-stage concept and technology development activities.[40]

General Findings Applying the CTP

We developed a CTP model to represent the components of the CTP. In using the model to analyze the CTP, its subordinate functions, and a subset of defense innovation organizations, we made a series of findings:

- There is no single "pipeline" or pathway of technologies through the CTP. There are many on- and off-ramps at every phase and feedback loops both within and between CTP phases. The CTP has many potential paths (combinations of activities within and between phases) from concept to fielding, and they are not necessarily linear or sequential.
- CTP functions are distributed across multiple stakeholders; no single organization performs them all. For example, defense innovation organizations perform multiple CTP functions and use a range of approaches. Most DIOs tend to focus on activities or functions in the identification or early-stage concept and technology development phases. Other CTP stakeholders also tend to focus on specific functions (program management, test and evaluation, etc.).
- Different combinations of organizations are involved in different phases. Given the variation in roles, responsibilities, functions, and activities among stakeholders, there are many opportunities for collaboration among organizations to move a technology forward.
- Most DIOs perform multiple functions to identify promising technology. Many innovation organizations seek to be guided by high-priority end-user problems and devote considerable time and effort to identifying and curating problems from across the department. Solution concept generation occurs in many ways that engage the ultimate customer or end user to a greater or lesser extent.
- All DIOs support the maturation of technology, but they often focus on different stages or means. DIOs have limited ability to facilitate the adoption of new technology in the final stages: the actual procurement and fielding of technology. In many cases, the adoption of the technology relies on finding a program or other customer to provide funding for adoption.
- DIOs' coverage of enabling functions is limited. Some DIOs perform navigation and coordination functions to help businesses understand and gain entry to the DoD market.
- No single organization has visibility into CTP activities enterprisewide. DIOs and other CTP stakeholders have visibility into their own activities, but they have limited visibility into the activities of other organizations. There is no oversight body or database that

---

[40] As noted above, RRTO is an exception.

tracks CTP-related activities across the DoD enterprise. Knowledge of other stakeholders tends to be ad hoc and driven by established personal or professional relationships among individuals.

- No DoD organization has responsibility and accountability for the performance of the CTP. No single organization "owns" the CTP as a process, and therefore no organization is solely responsible for outcomes. This may not be surprising, given that collaboration across multiple stakeholders is generally required to move an idea from concept to fielding, but it does make the CTP more difficult to govern in a way that will improve outcomes (accelerate commercial technology adoption and fielding for military use).

The fact that no single organization performs all necessary functions, and that there are many possible paths for taking a technology from concept to fielding, suggests that collaboration among stakeholders is required to accelerate the identification, development, and adoption of commercial technology for military use. Additionally, the CTP model suggests that CTP enabling functions guide, support, and drive activities in each of the three phases, including collaboration among stakeholders.

A top-down approach where the government tries to direct this system is likely self-defeating. We hypothesize, however, that throughput and effectiveness can be enhanced by policy levers that encourage and incentivize coordination and collaboration, striking a balance between organizational independence, free-market style competition and collaboration, and more centralized direction of the CTP. Relatedly, we hypothesize that different kinds of collaboration should occur, depending on the specific set of functions and activities a combination of stakeholder organizations perform. We examine these and other challenges and alternatives in Chapters 4 and 5, drawing on the discussion of the characteristics of a well-functioning CTP as presented in this chapter.

CHAPTER 4

Challenges and Gaps

With the CTP model (Figure 3.1) and the characteristics of a well-functioning CTP (Table 3.2) as reference points, we sought to understand how CTP functions are being executed and integrated by defense innovation organizations and other CTP stakeholders. In this chapter, we discuss the challenges and gaps that we identified. We found that each of the characteristics of a functioning CTP is lacking. These gaps and their impacts are summarized in Table 4.1. In the second half of the chapter, we examine gaps and challenges that are tied to particular CTP functions. Table 4.2 summarizes these issues and their impacts.

TABLE 4.1

Challenges and Gaps in CTP Characteristics

Desired CTP Characteristics	Challenge/Gap	Impact
Shared mission	CTP stakeholders are not aligned to a shared mission	Impedes technology transition and contributes to the "valley of death" problem
Common goals, objectives, outcomes	DoD has not established goals, objectives, or desired outcomes to guide CTP stakeholders	Stakeholder priorities and activities are not aligned, and collaboration is not incentivized, leading to inefficiencies and unnecessary costs
Clearly defined roles and responsibilities	Stakeholder roles and responsibilities are not well defined or understood	Creates confusion about who does what; leads to overlap across innovation organization activities and programming; leaves significant gaps in development activities and adoption functions
No gaps in key functions	*Gaps in key functions detailed in Table 4.2*	*Impact of these gaps detailed in Table 4.2*
Regular information sharing, coordination, and collaboration	Information sharing, coordination, collaboration, and feedback mechanisms between CTP stakeholders and functions are not established or functioning well	DoD does not realize value of problem curation, technology scouting, and other investments; CTP stakeholders do not know what others are focusing on, which impedes collaboration and contributes to the valley of death problem

Table 4.1—Continued

Desired CTP Characteristics	Challenge/Gap	Impact
Incentive structure aligned to CTP goals, objectives, and outcomes	Incentive structures for innovation organizations are not aligned to CTP goals, objectives, and outcomes	Key stakeholders lack incentives to collaborate and facilitate adoption of commercial technology; DIO-led problem curation, technology scouting, and early technology development are divorced from the traditional requirements and acquisition processes, contributing to the valley of death problem
Incentive structure aligned to CTP goals, objectives, and outcomes	No DoD-wide metrics and accountability mechanisms to check progress against common goals, objectives, and outcomes	DoD unable to monitor and assess progress against goals; unclear how to evaluate return on investment

Challenges and Gaps in CTP Characteristics

"The pipeline is broken because of incentive structures, a lack of unified DoD strategy, and mission [for innovation]."[1]

CTP Stakeholders Are Not Aligned to a Shared Mission

In a well-functioning CTP, stakeholders possess the shared mission of pushing promising technologies all the way through the pipeline. However, we found that CTP stakeholders are not aligned to a shared mission.[2] DIOs have disparate missions, and while many view themselves as playing a role in the identification, development, and adoption of commercial technologies, they do not view themselves as part of an integrated system. Indeed, different organizations and stakeholders lack a shared understanding of what constitutes innovation.[3] One interviewee noted that while Congress has regularly directed the Defense Department to adopt commercial technologies, Congress has not specified how the DoD should implement or enforce its direction, "so it doesn't get enforced."[4] We found that other key stakeholders in the traditional acquisition, requirements, budget, and end-user communities do not see themselves as part of the CTP at all, and DIOs lack consistent buy-in from the traditional communities. The lack of a shared mission and buy-in from traditional communities ultimately impedes technology transition and contributes to the "valley of death" problem.

---

[1] Interview 23, Interviewee 29, September 20, 2021.

[2] Interview 11, Interviewee 12, August 25, 2021. "There isn't a DoD-wide overall strategy specifying 'this is your specific mission set moving forward,'" the interviewee said. "Many are new organizations, and the mission hasn't been set yet."

[3] Interview 23, Interviewee 29, September 20, 2021.

[4] Interview 37, Interviewee 48, October 25, 2021.

DoD Has Not Established Goals, Objectives, or Desired Outcomes to Guide CTP Stakeholders

Contrary to what we would expect to find in a well-functioning CTP, we found that DoD has not established goals, objectives, and desired outcomes to guide CTP stakeholders. There is no DoD-wide strategy or policy guidance for innovation, and other than broad statements of modernization priorities, no specific goals and objectives have been articulated for the CTP.[5] Stakeholders throughout the enterprise each set their own priorities and objectives, which for the most part are not integrated or aligned. As one interviewee noted:

> We start with a deliberately balkanized approach on how we think about innovation. The services have autonomy and agency; OSD is not the be-all and end-all. From the word "go," we're talking five different approaches to innovation [from the services], maybe six when you bring in OSD trying to create fourth estate value. We start with a disparate approach. Even in the services there are different pieces. The acquisition executive is not the same as the CTO, [who] is not the same as the chief innovation officer, [who] is not the same as the four-star officer in charge of requirements. None reports to a common, single person who presents a common picture of what we mean when we say "innovation approach."[6]

While this "let a thousand flowers bloom" approach has some theoretical upsides—for example, maximizing autonomy, flexibility, and speed—in practice, the diversity of definitions, approaches, and priorities across DoD and distribution of funds across innovation organizations and other CTP stakeholders means that stakeholder activities are not aligned, and collaboration is not occurring. Additionally, the fragmented approach to innovation leads efforts to identify common problems and technologies to be duplicated across DIOs and military services, leading to inefficiencies and unnecessary costs for the government and industry partners.[7] One interviewee noted that "[without] direction from OSD [to set priorities,] I feel like we are chasing every shiny penny across the department and the services. Our industry partners might have technology to sell to all of us, but the work is multiplied three-fold for them between the Army, the Navy, the Air Force, and the Department of Defense. I don't think it's helpful when we're all pursuing our own end states, without any kind of strategic approach. Maybe there is a strategic approach, and I don't know what it is."[8]

---

[5] Interview 11, Interviewee 12, August 25, 2021; Interview 18, Interviewee 20, September 15, 2021; Interview 23, Interviewee 29, September 20, 2021. Noting the lack of policy guidance and regulations for the innovation ecosystem, one interviewee noted that "someone asked about regulations, and I said there isn't one. I was told there a regulation on everything [in DoD]: there are top-level regulations and then office-level regulations derived from the top one. But there is nothing like that [for innovation]." Interview 7, Interviewee 8, August 11, 2021.

[6] Interview 23, Interviewee 29, September 20, 2021.

[7] Interview 18, Interviewee 20, September 15, 2021.

[8] Interview 18, Interviewee 20, September 15, 2021.

Stakeholder Roles and Responsibilities Are Not Well Defined or Understood

We found that CTP stakeholder roles and responsibilities are not well defined or understood. Within the innovation ecosystem, there are a large number of innovation organizations—we counted more than 60 but expect that we did not identify all of them (others have put the number at over 100)—and neither prospective partners in the government nor new-entrant businesses know that they exist or understand what they do. As an interviewee said, "One concern raised to us from the commercial sector is, who do I reach out to for what? There is some confusion. As part of the U.S. government, how many innovation organizations including R&D are there that are foundational to advanced technologies? Over 100. For small companies, that becomes very confusing. . . . It gets complicated very quickly."[9]

The problem persists even within the department. One DoD innovation organization representative spends a lot of time establishing relationships within the department, helping DoD entities understand what the innovation organization is, building trust, and getting others to understand that the organization is an official DoD entity. The representative noted that he was often met with skepticism: "I've had people in the [services] think I'm a spy based on some of the questions I've asked. . . . New people who have joined a prospective DoD customer are the hardest to break in."[10]

There appears to be considerable overlap across innovation organization activities and programming. Many DIOs concentrate on early CTP functions (the "identification" bin), such as conducting acceleration programs and offering support to develop solution concepts, and then prioritize commercial technologies in defined TRL ranges. For example, many DIOs favor SBIR as a principal contracting vehicle for follow-on activities with commercial businesses. Phase I awards cover basic research and experimentation across TRLs 1–3, while phase II awards cover prototyping and demonstration across TRLs 4–6. Although some DIOs do occasionally award some phase III contracts, the focus of the DIOs (through SBIR and other programming) is concentrated in the early-stage (TRLs 1–6) projects.[11]

We also noted significant gaps in development and adoption functions. We identified many examples of support for early-stage concept and technology development, but support and opportunities for more advanced development activities such as prototypes, live exer-

---

[9] Interview 11, Interviewee 12, August 25, 2021.

[10] Interview 4, Interviewee 5, July 20, 2021.

[11] DoD acquisition policy states that TRL 6 is the technology maturity threshold required for a system to clear milestone B and enter the engineering and manufacturing development phase for a major capability acquisition (see DoDI 5000.02 and DoDI 5000.85). However, not all TRL 6 ratings are equal. The technologies in a major capability program in the traditional process go through a technology maturation and risk-reduction phase leading up to a milestone B decision that is likely much more intensive than a prototyping demonstration activity that a DIO can fund.

cises, and demonstrations are distributed unevenly across the department.[12] Conversely, traditional acquisition stakeholders, including requirements developers, PMs, PEOs, and end users, view their principal responsibilities narrowly and do not appear to view themselves as responsible for the integration of dual-use technologies into the department.

Information Sharing, Coordination, Collaboration, and Feedback Mechanisms Between CTP Stakeholders and Functions Are Not Established or Functioning Well

Given that no single organization performs all the key functions required to effectively accelerate the identification, development, and adoption of technology for military use, linkages between defense innovation organizations are essential. However, we heard from many interviewees that these linkages do not appear to function well, and innovation organizations and data are stovepiped: most innovation organizations do not share information and do not coordinate or collaborate routinely with other organizations.[13]

There are no formal mechanisms or requirements to share information such as DoD problems or potential commercial solutions or to coordinate or collaborate across the defense innovation ecosystem.[14] One interviewee noted, "Currently there is no mechanism or incentive for different service components to talk to one another. There is no organizing function for them to do it. . . . OSD/R&E [Office of the Secretary of Defense, Research and Engineering] is focused on modernization priorities, but there is no convening function that exists for innovation organizations in the context of R&E."[15]

Some organizations seek to identify "transition partners" with whom to collaborate to lay the groundwork for later handoffs, but we observed that for a variety of reasons, there is often little follow-though by those partners, and technologies are rarely fielded at scale. As described in greater detail below, we found that warm handoffs between stakeholders across the CTP are generally not occurring, and feedback mechanisms between CTP functions are not formally established. Notably, we found that DIOs and traditional acquisition stakeholders are not coordinating or collaborating effectively to develop and transition technologies to fielding. According to one interviewee, most DIOs "have no real sense of what the actual capability requirements are because almost every innovation organization does not talk to

---

[12] The DoD Rapid Reaction Technology Office is an example of an organization that supports these activities. CP/RRTO is discussed in greater detail in the Appendix A.

[13] Interview 1, Interviewee 1, July 20, 2021; Interview 21, Interviewee 25, September 16, 2021; Interview 27, Interviewee 34, September 24, 2021. "The biggest issue is we shouldn't be studying something that someone else has already studied, but we don't know it. I try to share what we're working on with other centers [within my organization]. The key is to talk to those content authors who have already worked on aspects of similar problems, [but] there is a lot of stovepiping of data." Interview 34, Interviewee 45, September 29, 2021.

[14] Interview 11, Interviewee 12, August 25, 2021.

[15] Interview 30, Interviewee 40, September 24, 2021.

PMs or PEOs."[16] The upshot of this disconnect is that the DIOs "have lost the bubble on where decisions are being made."[17]

When information sharing, coordination, and collaboration occur between stakeholders, these activities are ad hoc and often based on personal relationships.[18] One innovation organization representative noted that "my one-word job description is transparency. For example, if [my innovation organization] is spending millions on a helmet, and others use helmets, I want to make sure that information gets shared and all the resources can be shared so we are not duplicating efforts."[19] Some organizations seek input from others on the technologies that they are considering supporting—CP/RRTO is one example—but this is not the norm and appears to be a mostly one-way conversation.

As a result, DIOs appear to have limited insight into what other DIOs are focused on and the problems and related customers they serve. Each DIO takes its own approach to developing its own ecosystem—through market scanning and developing ties to academia, research organizations, and start-ups—and they only share this information haphazardly. This leads to inefficiencies across the innovation ecosystem. One interviewee noted, "The Department has an 'innovation archipelago.' Many offices within DoD are engaged in excellent and important work on innovation, but each is an island, disconnected from the rest. This lack of communication and collaboration is hampering progress."[20]

Another interviewee said that his office tries to promote service innovation opportunities to small businesses through the office's social media account, website, and in engagements with small businesses, but noted that a lack of information about those opportunities prevents his office from doing more:

> Nobody knows what anyone is doing. There is a lack of information flow. I can't fully perform my responsibilities because I don't know. There's a vacuum of knowledge. Everybody is out there doing—which is important—but I think if there was a formal requirement for collaboration, then we would have better access to that knowledge and better be able to have a common approach, message within industry. . . . My office is continually

---

[16] Interview 23, Interviewee 29, September 20, 2021.

[17] Interview 23, Interviewee 29, September 20, 2021.

[18] Interview 35, Interviewee 46, October 19, 2021. According to this interviewee, "A couple years ago, the OSD/A&S [Office of the Secretary of Defense, Acquisition and Sustainment] started a weekly sync between DoD innovation programs that was run out of the industrial policy office. That's the same group that started the trusted capital marketplace. They had a keen interest in seeing what all the innovation programs were doing, companies being selected, and bring[ing] them into the trusted marketplace. Informally, the first people I reached out to were AFWERX and SOFWERX. Through them, like a telephone game, we got introduced to others and others. Those relationships were personal mainly, so they ebbed and flowed. There was not enough formal coordination across these programs. More recently, through the trusted capital working groups we've talked a lot more about data sharing. Prior, a friend at SOFWERX would call and ask, 'Hey, have you seen technologies on counter-drones?' I'd email them a listing. That's not a sustainable way to do it."

[19] Interview 26, Interviewee 33, September 24, 2021.

[20] Interview 11, Interviewee 12, August 25, 2021.

playing catch-up, chasing after what everyone is doing, so we can push things to industry. I don't think there is a widespread understanding among small business professionals about what incubators and accelerators are and how to partner with them to help industry to better engage with government pockets of opportunities. . . . What a missed opportunity for us not to know about these things and to help push them. We literally do thousands of engagements across [the service], multiple across DoD. We are going to thousands of engagements [with small businesses], promoting these opportunities. Not having that strategic approach to small business function and innovation is a missed opportunity. There are too many entrances to work with.[21]

Incentive Structures for Innovation Organizations Are Not Aligned to CTP Goals, Objectives, and Outcomes

"I tell generals and SESs [members of the Senior Executive Service] that you can't stand up [an innovation ecosystem] and tell people to think about using it. You have to say 'thou shalt use this tool in your toolbelt to help.'"[22]

Incentive structures for innovation organizations are not aligned to produce CTP outcomes. Key stakeholders lack incentives to collaborate and to facilitate adoption (fielding) of DIO-identified commercial technologies. Program managers are often reluctant to integrate early-stage technologies in their programs, including pursuing new development efforts in parallel, because of added risk to the program. This reluctance principally arises because PMs typically defer to the prime contractor on component or subsystem integration to minimize risk to the government.

Problem curation and technology development tend to be divorced from the traditional requirements process. Without a requirement, program managers are reluctant to support those technologies. Although many DIOs seek end-user feedback and buy-in on technologies entering the pipeline, end users often do not have direct influence over investment or technology decisions.[23] No other stakeholders are filling these gaps. The lack of incentives supporting CTP goals, objectives, and outcomes contributes to the valley of death problem, discussed below.

CTP incentives can be shaped by leaders and oversight bodies, but interviewees noted that there is no senior DoD advocate for innovation who is regularly checking on progress and asking tough questions:

There is no senior leader in the executive branch that cares about what innovation organizations are doing, which leaves innovation organizations with only one customer left to impress: Congress. So you try, three to four times a year, to do a briefing, show transfor-

---

[21] Interview 18, Interviewee 20, September 15, 2021.

[22] Interview 26, Interviewee 33, September 24, 2021.

[23] Interview 23, Interviewee 29, September 20, 2021.

mative projects, and come back with [requests for information] that you spin back to the committee. You're not really responsible for change in the department, you're responsible [for showing] you're doing something interesting to Congress. There is no incentive structure for innovation organizations to show how many of their claimed 12 "transformative" projects have transitioned to programs of record [and] how many are in the FYDP [Future Years Defense Program]. If the answer is none, what are we doing then? Congress doesn't have the time to do that deep dive, so no one is asking those hard questions.[24]

DoD Has Not Established Metrics and Accountability Mechanisms to Check Progress Against Common Goals, Objectives, and Outcomes

"[This is] innovation theater. . . . It's all 'flash, bang, sizzle.' The reality [is that what it really takes to help start-ups sell to DoD] is unsexy."[25]

A well-functioning CTP should be buttressed by metrics and accountability mechanisms to check progress against preestablished goals, objectives, and outcomes. We found that these metrics and mechanisms are currently lacking. Where they do exist, metrics and mechanisms are focused mostly on outputs for a DIO, such as the number of solicitations posted, proposals received, early-stage venture participants in DIO-funded accelerators, and prototype projects initiated.[26]

DIOs have encountered impediments to developing meaningful metrics that link activities and outputs to outcomes.[27] One interviewee noted that "overall impact [is] harder to measure. With respect to cost savings, saving lives, operational efficiency—those are metrics that are hard to track."[28] Another interviewee said, "It's hard to define what success is. . . . The problem is, I'm a lowly innovation institute, hamstrung by bureaucrats who control the purse strings. Somewhere within there, someone will get credit for something my shop did, and it won't be me."[29] Compounding this issue, CTP stakeholders have no common definition of transition, which many organizations identify as a key metric. For some, "transition" means transition to another program within the organization or to another DIO.[30] For others, it means fielding to any DoD customer—even a single unit—or knowledge transfer.

Reflecting on innovation metrics, one interviewee said, "What I tell all partners and collaborators [is] if participation is your only metric, you're running innovation theater. If you

---

[24] Interview 23, Interviewee 29, September 20, 2021.

[25] Interview 30, Interviewee 40, September 24, 2021.

[26] Interview 11, Interviewee 12, August 25, 2021; Interview 29, Interviewee 39, September 29, 2021.

[27] Interview 1, Interviewee 2, July 20, 2021; Interview 11, Interviewee 12, August 25, 2021.

[28] Interview 11, Interviewee 12, August 25, 2021.

[29] Interview 29, Interviewee 39, September 29, 2021.

[30] Interview 11, Interviewee 12, August 25, 2021; Interview 10, Interviewee 11, August 2, 2021.

don't have transitions and knowledge transfers, you're likely doing innovation theater."[31] Another interviewee said, "[My organization] is seeking to dazzle its shareholders—Congress—with stories of success. To do this, we measure what we're doing for customers to justify our funding. For example, we tell dazzling stories of students developing a technology that DoD wants, but I question whether [interaction with my innovation organization] was necessary to achieve that end."[32]

We explored "success" stories of companies that have promising technologies and have worked with one or more DIOs (see Appendix B). Almost none of these technologies have been fielded to a military user. From the DIOs' perspectives, these interactions with businesses were successes because the DIOs supported development of innovative technologies and acculturated the businesses to contracting with the government. From the businesses' perspectives, their interactions with DIOs were limited successes, helping them overcome some challenges and continuing to develop their technologies. However, most of the businesses, with the exception of Company X (a business that asked for anonymity) and Vita Inclinata, noted that they have not yet sold to the department. Whether participation in DIO programs ultimately helped the businesses establish long-term dual-use viability or become fully integrated into DoD's supply chain therefore remains to be seen.

Finally, from DoD's perspective, it is unclear how to evaluate return on investment in these cases. Should the creation of a new entrant to the DoD innovation ecosystem be considered a successful outcome? Perhaps it makes sense to consider different "tiers" of success, from limited outputs at the DIO level to transition rate and contribution to operational effectiveness at the enterprise level. However, DoD has not defined success or specified how to measure progress, leading to disjointed and inconsistent standards across the ecosystem.

Challenges and Gaps in CTP Functions

We turn now to gaps and challenges that are tied to particular CTP functions. Table 4.2 summarizes these issues and their impacts. Note that some of the CTP functions are grouped together because they relate to the same challenge and impact.

Identification of Commercial Businesses and Technologies Appears Unsystematic

Innovation organizations have no shortage of interesting technologies, but identification of commercial businesses and technologies appears unsystematic. Because there is no comprehensive approach in use to identify businesses or technologies, DoD may be overlooking promising technologies.

This gap ties to an identification function: technology scouting.

---

[31] Interview 26, Interviewee 33, September 24, 2021.

[32] Interview 5, Interviewee 6, August 5, 2021.

TABLE 4.2

Challenges and Gaps in CTP Functions

Challenge/Gap	Impact	CTP Function
Identification of commercial businesses and technologies appears unsystematic	DoD may be overlooking promising technologies	Technology scouting
There is no institutionalized approach to outreach or enterprise solution to store and share data	Contributes to duplication of effort, inefficiencies, and missed opportunities; DoD not capturing the full value of labor-intensive outreach and networking with commercial businesses and DoD partners	Network development
DoD end users do not effectively communicate capability needs	Contributes to duplication of effort and inefficiencies (as users seek to solve related problems independently) and missed opportunities to target novel solutions to problems	Problem identification, validation, and curation
Problems are not shared across CTP stakeholders	Contributes to duplication of effort, inefficiencies, and missed opportunities to target novel solutions to problems; DoD not capturing the full value of labor-intensive problem curation	Problem identification, validation, and curation; solution concept generation; customer discovery; matching problems and solutions
Businesses cannot easily locate an initial point of entry into DoD	Industry unable to identify and access DoD opportunities and understand how their technology applies to DoD missions	Technology scouting, customer discovery, solution concept generation
Businesses find it challenging to understand and "speak" DoD	Industry does not understand what DoD needs or how their technology may be relevant	Customer discovery, solution concept generation, and (indirectly) development and adoption functions[a]
Businesses cannot easily identify the right DoD officials to provide feedback on their technology	Businesses lack feedback that is required to refine a technology's application within a mission space and develop and mature promising technologies	Customer discovery, solution concept generation, matching problems and solutions, and all development functions
There are no warm handoffs, clear routing, or obvious next steps for businesses	Businesses struggle to understand what is next and to identify subsequent funding and support; technology development and adoption may stall	All development and adoption functions
Suitable funding for continued development of technology is limited, and available funding is difficult to locate	Leads to gaps in funding and delays; technology development may stall; business may struggle to remain viable	All development and adoption functions
Most existing opportunities are overly burdensome to pursue	Burdensome processes are disproportionately costly for new entrants, deterring them from working with DoD	All development and adoption functions

Table 4.2—Continued

Challenge/Gap	Impact	CTP Function
Many factors (including funding, misalignment with budget process, technical failure, and lack of a formal requirement or customer who is ready) contribute to the "valley of death"	Efforts to address the valley of death may fail because they focus on only one of these factors	Transition to fielding
The valley of death persists in part because no DoD organization has visibility into all CTP activities or responsibility for CTP outcomes; gaps between various stakeholders hinder adoption	Technologies stall in the valley of death without a clear path to continued development or transition	Transition to fielding

[a] This gap ties to multiple development functions, including further development and maturation of solution concept, S&T activities, early-stage technology maturation, RDT&E activities, and late-stage product maturation.

There Is No Institutionalized Approach to Outreach or Enterprise Solution to Store and Share Data

In seeking to identify technologies, several of the DIOs that we spoke with noted that they view expanding their networks as an explicit part of their work, including the need to identify new entrepreneurs, research institutions, and academics working in identified problem areas.[33] However, many innovation organizations primarily rely on personal networks. Without an institutionalized approach to outreach and networking and an enterprise solution to store related data, such as enterprise CRM software, this approach renders the organization vulnerable to turnover.[34] One interviewee noted that her organization's network is entirely dependent on personal relationships, and when one node—either the DoD contact or the innovation organization representative—changes, the relationship is at risk.[35] Another interviewee said, "We have a lot of solutions and exposure to the department, but we are not leveraging our network."[36] Others noted that their organizations did not yet know the right people.[37]

---

[33] Interview 4, Interviewee 5, July 20, 2021.

[34] Interview 1, Interviewee 1 and 2, July 20, 2021; Interview 35, Interviewee 46, October 25, 2021; Interview 17, Interviewee 19, September 24, 2021.

[35] "Personal relationships are great and core, but is that something that is scalable?" Interview 1, Interviewee 1, July 20, 2021.

[36] Interview 1, Interviewee 2, July 20, 2021.

[37] Interview 31, Interviewee 42, October 6, 2021; Interview 9, Interviewee 10, August 18, 2021; Interview 21, Interviewee 25, September 16, 2021; Interview 29, Interviewee 39, September 29, 2021.

Some innovation organizations rely on intermediaries like venture capitalists and industry consortia to help identify companies,[38] but these shortcuts also have limitations. For example, consortia managers search for technologies within the consortia in response to a specific DoD requirement but have few incentives to proactively search for new technologies and recruit new members into the community. This approach, on its own, may cause innovative technologies to remain unknown to, or overlooked by, DoD.

These gaps contribute to duplication of effort, inefficiencies, and missed opportunities. DoD is not capturing the full value of labor-intensive outreach and networking with commercial businesses and DoD partners.

This gap ties to an identification function: technology scouting.

DoD End Users Do Not Effectively Communicate Capability Needs

DIOs face a number of challenges in seeking to identify and curate problems from across the department. Representatives of some DIOs noted with frustration that end users and other stakeholders within the same organization are not good at articulating capability needs to innovators.[39] These stakeholders identify different priorities, and those priorities shift, which has meant in some instances that a technology identified and developed by a DIO is no longer of interest to the end user.[40] Innovation organizations reported that operational units are not aware of other units' problems and are seeking to solve them independently through different channels: "All the problem statements, there's not a lot of connective tissue between units. We'll see four to five people trying to solve the same problem."[41] During the acquisition policy game, we observed that several DIOs were willing to depart from end users' stated preferences to pursue technologies that the DIO identified as "interesting."[42] For their part, end users may not be well-postured to evaluate whether early-stage technologies can help them.[43]

These challenges contribute to duplication of effort and inefficiencies, as users seek to solve related problems independently, and missed opportunities to target novel solutions to problems.

These challenges tie to an identification function: problem identification, validation, and curation.

Problems Are Not Shared Across CTP Stakeholders

From a CTP perspective, the innovation organizations that we examined appear to have their own approach to sourcing, binning, and cataloging problems. This work is challenging and

---

[38] Interview 1, Interviewee 2, July 20, 2021; Interview 12, Interviewee 14, August 26, 2021.

[39] Interview 35, Interviewee 46 October 19, 2021.

[40] Remarks from participant in acquisition policy game, March 9–10, 2022.

[41] Interview 9, Interviewee 10, August 18, 2021.

[42] Remarks from participant in acquisition policy game, March 9–10, 2022.

[43] Remarks from participant in acquisition policy game, March 9–10, 2022.

human capital–intensive, and the resulting set of problems is a potentially valuable output of DIOs' work. However, DIOs do not have a formal or regular mechanism to document and share problems with other innovation organizations or DoD.[44] One innovation organization representative noted that problem identification and curation is one of her organization's most valuable outputs, but the innovation organization does not have any approach or mechanism to bin or catalog problems or share problems even within her organization.[45] This may result in duplication of effort within and across organizations, mismatches between the problem owner's needs and the available DIO capabilities, and missed opportunities to target novel solutions to problems. If there is no immediate match between a problem and possible solution, the problem, end user, and/or solution may be abandoned.[46] This means that DoD is not capturing the full value of labor-intensive problem curation.

This gap ties to multiple identification functions: problem identification, validation, and curation; solution concept generation; customer discovery; and matching problems and solutions.

Businesses Cannot Easily Locate an Initial Point of Entry into DoD

For those businesses without a background in DoD and with limited knowledge of DoD organizational structure and processes, there is no centralized, easily navigable entry point where businesses can identify and access DoD opportunities and understand how their technology may apply to DoD missions. One interviewee noted, "There's no entry door for commercial companies to DoD, and that's a big problem. They have a good idea of who to go to [on the commercial side]—the CTO if they're selling IT stuff, the head of certain divisions—but the military is just a black hole. All these operational heads without any procurement authority, they don't know where to start."[47] Another interviewee noted that "DoD is not organized to set up one or even four doors to entry, so that if you're on the outside trying to get into this gray monolith, we haven't set up one or five doors, we have 105. You can't find your way into the maze in the right way. DoD is deeply unorganized. What we invite is frustration from those the DoD needs. We say, 'look we're trying to be agile!' and a company walks in to find out they aren't in the right door."[48] As a result, industry is unable to identify and access DoD opportunities and understand how their technology applies to DoD missions.

The new generation of defense innovation organizations was created in part to help solve this problem—to help give new-entrant companies a place to start. But given the proliferation of these organizations, companies still do not know which door to enter.[49] OSD recently

---

[44] Interview 4, Interviewee 5, July 20, 2021.

[45] Interview 4, Interviewee 5, July 20, 2021.

[46] Interview 12, Interviewee 14, August 26, 2021.

[47] Interview 9, Interviewee 10, August 18, 2021.

[48] Interview 23, Interviewee 29, September 20, 2021.

[49] Interview 9, Interviewee 10, August 18, 2021.

sought to address this gap, creating an "Innovation Pathways" website that is intended to provide a "'one-stop shop' for the DoD innovation ecosystem."[50] The website collates links to innovation organizations and opportunities and organizes them according to the prospective user. There are tabs for innovation organizations, business and industry, students, universities, and DoD military and civilian users. At the time this report was drafted (in May 2022), the website did not include information relating to these linked organizations and opportunities. To find more information, users are required to identify and click on links to other organizational websites. The search function is extremely limited, allowing a user to search only the information visible on the page that is currently displayed. As of September 2022, users cannot do a keyword search across the Innovation Pathways website of the Office of the Under Secretary of Defense, Research and Engineering (OUSD/R&E) or its subordinate links. A search for "SBIR," "hacking," "accelerator," "development," and "funding" produced no results. The website marks progress and can be improved over time, but at present, interested businesses and other users will still be required to click through the site and navigate a large number of other websites to identify potentially relevant organizations and opportunities.

As a result, DoD is not reaching all companies developing innovative technology with potential military applications and may be missing many that could be working with DoD. Most interviewees who discussed this indicated that it was a significant drag on bringing new entrants into DoD, though other interviewees downplayed this issue:[51] "I think we are missing a whole bunch. I think it's a problem to be solved. I don't think we're doing a particularly good job finding innovators, or finding companies that have technologies that we would be interested in. That does mean we have to be going into spaces where those companies go. I don't think we have a good understanding of what all the industry are."[52]

Several interviewees from both DIOs and new-entrant businesses made the point that being physically present where innovation ecosystems concentrate is important.[53] The commercial sector creates informal and formal networks of entrepreneurs, venture capital, and more mature businesses colocated in geographic regions through frequent in-person engagements. DoD needs to engage businesses in these environments. As one interviewee from a new-entrant company noted:

> I didn't need handholding [from DoD]. No good entrepreneur needs handholding. But I need someone to point me in the right direction, tell me [the department's] priorities, and I'll figure it out. Those things are hard to do if the institution doesn't have a presence in the Silicon Valley. What DoD can do differently is, first, to have way more people in the Valley. Physical presence is important: They need to be able to walk back, do active

---

[50] U.S. Department of Defense, "Defense Department New Website to Navigate Innovation Opportunities," press release, April 22, 2022; OUSD/R&D, undated a; Office of the Under Secretary of Defense, Research and Engineering, "Innovation Pathways," homepage, undated b.

[51] Interview 12, Interviewee 14, August 26, 2021.

[52] Interview 18, Interviewee 20, September 15, 2021.

[53] Interview 4, Interviewee 5, July 20, 2021; Interview 42, Interviewee 54, November 10, 2021.

technology scouting, go out and meet with start-ups virtually or in-person to figure out what's going on and who's doing what. There is so much sclerosis. For a long time, the Army and Air Force thought if there was a good opportunity, it would come to them. They have learned that that's not the case; they have to go to the Valley now. As an example, automotive companies work out of Michigan, but they have a large presence in the Valley. It doesn't have to be physical, but this connection needs to happen. It's networking, attending events, telling the start-up community that [your organization] is open for business. Programs like xTechSearch—they're really cheap start accelerators: $10,000 on offer. There are raw start-ups paying no attention to DoD, but if they know they can get $10,000 for just giving a presentation and an idea, that will interest them.[54]

That argues for the regional hub presence that some DIOs maintain.

After businesses and technologies have been identified, DIOs have no formalized mechanisms to share information relating to interesting technologies with other CTP stakeholders.[55]

This gap ties to identification functions: technology scouting, customer discovery, and solution concept generation.

Businesses Find It Challenging to Understand and "Speak" DoD

"How to talk to [DoD], where to talk to [DoD], and when to talk to [DoD] are huge barriers."[56]

Businesses find it challenging to understand and "speak" DoD. Discussing one early-stage venture, an innovation organization program manager noted: "The founders are smart; they have the technology chops. And like any start-up, they didn't know how to speak DoD. . . . Early-stage ventures like these need to find shortcuts and ways to understand DoD, routes in, to test the technology, and to speak DoD. They need to learn where to find this information."[57] A service representative echoed these sentiments: "Industry doesn't know where to go, and they're the ones who feel the impact. [Within the government] we can work within ourselves and figure it out. The industry doesn't understand us—we're big. The Army is big, let alone the Department of Defense. [Leadership] has everybody going out trying to find the next latest and greatest things, but early-stage ventures don't know how to pitch. who are the right people [in the DoD] they should be talking to? We don't have a unified management approach on how to [help them]."[58]

---

[54] Interview 42, Interviewee 54, November 10, 2021.

[55] Some interviewees reported informal channels through which they reach out to other CTP stakeholders, including program managers and end users, but the channels that were described to us were ad hoc and relationship-dependent.

[56] Interview 18, Interviewee 20, September 15, 2021.

[57] Interview 30, Interviewee 40, September 24, 2021.

[58] Interview 18, Interviewee 20, September 15, 2021.

DIOs also have difficulty distinguishing between two different types of "end users": requirements developers and operational users. Engagement with the right kind of DoD customer is critical to refining technology or product solutions to address a real capability need. Businesses new to DoD do not understand the scope of authority, or the role, of a given customer. They may not recognize the difference between requirements developers and operational users, and they may find the requirements, acquisition, budgeting, and decision processes difficult to understand and align to their own needs for financing or access to test facilities.

As a result, industry does not understand what DoD needs or how its technology may be relevant.

This challenge ties to customer discovery and solution concept generation and, indirectly, to development and adoption functions.

Businesses Cannot Easily Identify the Right DoD Officials to Provide Feedback on Their Technology

Identification of prospective customers and end users is also essential for businesses to gain access to prospective DoD customers to understand their end-user needs.[59] Direct engagement with prospective end users helps businesses and DIOs understand potential uses of their technology and how to frame and tweak their technology to build a product that will be useful to DoD.[60] Without this input and engagement, businesses and DIOs may not fully understand how the technology could help and how to refine the technology's application within a mission space. As we have discussed, however, businesses have difficulty identifying and becoming known to prospective customers, end users, and acquisition professionals within DoD.

One company we spoke with, for example, started its engagement with DoD by targeting service labs, believing the labs would be most interested in discovering and developing new technology. Yet the company realized that these labs were not well positioned to take advantage of technology developed in the private sector. This led the company to seek out assistance from a variety of DIOs.[61] Another company started by military veterans relied on existing relationships and knowledge of DoD processes to identify entry points, and it still faced challenges gaining attention.[62]

These engagements are also relevant later in the process. Maturation of the solution concept is a significant step for innovators, moving from the early-stage activities to actual capability development activities. Companies need to engage prime contractors, DoD stake-

---

[59] "We don't care about the money so much—we want an end user. We want a customer to say yes rather than some high-paying contract." Interview 20, Interviewee 22, September 15, 2021.

[60] One early-stage venture founder noted that they are seeking end users to serve as thought partners willing to "toss around ideas with us and exchange ideas." Interview 20, Interviewee 23, September 15, 2021.

[61] Interview 49, Interviewee 67, January 13, 2022.

[62] Interview 41, Interviewee 53, November 10, 2021.

holders, and regulators to understand DoD problems, to gain the insights needed to refine solution concepts, to develop and transition their technology, and to help inform proposals for related solutions, yet they may not know where to start.[63] Unlike traditional DoD vendors, early-stage ventures do not have touchpoints in the military services.[64] "Smaller businesses without traditional contacts with the [service] have a difficult time navigating the bureaucratic, policy, and legal aspects involved with [service] contracts but also getting a foot in the door to get feedback. It was a closed door for them without those connections," said one interviewee.[65] Another interviewee noted, "Trying to find the right program manager shop for your technology, that's half your battle."[66] It can also be difficult to convince these same entities that a cutting-edge technology is viable and could be integrated with existing technology.

This challenge ties to several identification functions: customer discovery, solution concept generation, matching problems and solutions, and all development functions.

There Are No Warm Handoffs, Clear Routing, or Obvious Next Steps for Businesses

For businesses, once they have come through a particular program or innovation organization, there are no obvious next steps.[67] Some handoffs exist within innovation organizations, program to program, but rarely between innovation organizations. Businesses struggle to understand what is next and to identify subsequent funding and support. Because they have no sustained assistance in identifying next steps, often technology development and adoption stall. One early-stage venture representative noted, "There's such a focus on getting small business into the defense ecosystem. Once they're in, it falls apart. We won [a service-led competition]. Hypothetically, this is the golden child of [service] procurement, and short of $250,000 in prize money, nothing else came of it."[68]

Representatives of businesses that we engaged suggested that sustained navigation or "concierge" support would be useful to help them navigate DoD. The defense innovation ecosystem is only partially addressing these challenges. Many of the DIOs do provide some initial guidance through accelerator and other similar programs but not through all CTP phases (identification, development, and adoption). After this initial touchpoint, many businesses need continued help navigating the DoD bureaucracy, policy, legal, and contracting aspects of technology development. Other CTP stakeholders—capability development/requirements and acquisition communities—do not provide navigation services.

---

[63] Interview 39, Interview 50, October 27, 2021.

[64] Interview 30, Interviewee 40, September 24, 2021.

[65] Interview 35, Interviewee 46 October 19, 2021.

[66] Interview 32, Interviewee 43, October 12, 2021.

[67] Interview 5, Interviewee 6, August 5, 2021; Interview 6, Interviewee 7, August 11, 2021; Interview 18, Interviewee 20, September 15, 2021.

[68] Interview 32, Interviewee 43, October 12, 2021.

This is a challenge for businesses and innovation organizations alike. One interviewee from an innovation organization noted that the lack of pathways and warm handoffs is "tough for all of my program managers. We're looking forward to getting to the passing zone, to hand the baton off, but heretofore, it hasn't materialized. . . . For [one of my] programs manager[s] in particular, this is debilitating. The work he does with the teams after they win is really keeping one hand tied behind his back in terms of capacity."[69]

This gap ties to multiple development functions, including further development and maturation of solution concept, S&T activities, early-stage technology maturation, RDT&E activities, and late-stage product maturation.

This challenge ties to all development and adoption functions.

Suitable Funding for Continued Development of Technology Is Limited, and Available Funding Is Difficult to Locate

Businesses that are not already familiar with DoD have a difficult time locating information about opportunities on public government websites and portals.[70] Many DIOs post opportunities on their webpages, relying on businesses to find them. They may also cross-post solicitations to SAM.gov (e.g., SOFWERX, AFWERX) or challenge.gov (e.g., xTech, Doolittle), but this is not consistent across the DIOs, and the information provided in the solicitations may not be enough for a small business to determine whether its technology is a good fit.

Many small businesses do not know that they can register and find contracting opportunities on SAM.gov, the official U.S. government system for contracting opportunities, or how to navigate the site or those opportunities.[71] Even well-publicized opportunities are confusing for new entrants to navigate. As one interviewee commented, "Deputy Secretary Hicks just started a Rapid Defense Experimentation Reserve fund to encourage joint collaboration and exercises. But alongside others—such as CP/RRTO, modernization efforts, NSIN, DIU, and so forth, it is still not clear . . . if I'm interested in hypersonics, where do I go? Who has the money?"[72] Another interviewee noted: "One thing government is terrible [at doing] is marketing because we haven't had to do it. Some think we put something on SAM.gov and the world will see it."[73]

We heard from many representatives of new-entrant businesses that there is limited funding available to develop promising technology that is suitable for a particular technology's

---

[69] Interview 5, Interviewee 6, August 5, 2021.

[70] Interview 2, Interviewee 2, July 30, 2021; Interview 9, Interviewee 10, August 18, 2021; Interview 11, Interviewee 12, August 25, 2021; Interview 18, Interviewee 20, September 15, 2021. Section 809 Panel, *Report of the Advisory Panel on Streamlining and Codifying Acquisition Regulations*, Vol. 1, January 2018, pp. 9–10.

[71] Sam.gov, System for Award Management, webpage, undated.

[72] Interview 30, Interviewee 40, September 24, 2021; Sydney J. Freedberg, Jr., "Hicks Seeks to Unify Service Experiments with New 'Raider' Fund," *Breaking Defense*, June 21, 2021.

[73] Interview 14, Interviewee 16, September 2, 2021.

maturity and application. One interviewee said that "the funding available through DoD is only for the product, not enough to sustain a company's hiring and processes. DoD funding is very minimal; most of the time, our costs far superseded money that is coming through. For early prototypes, it makes sense, but once it's a real thing and [there is] a proof-of-concept, [and] once you get into test and need to deploy, then it's not much compensation."[74] As another interviewee noted, it can be difficult to get those "first dollars" to mature a technology: "No one wanted to be first to take that chance on us. It was helpful to say we have been funded through [an innovation organization], but there is a piece missing because we haven't deployed anything. We don't have contacts [from the innovation organization with which we worked] who can say it works well. We're missing those endorsements."[75] Another interviewee noted: "We are at TRL [technology readiness level] 4. If we were Raytheon, we would be able to call [the DoD] and arrange a test of our technology. But we aren't Raytheon, and we don't think they would let us. We don't know how to get to the end user. Raytheon has a history of working from TRL 5–7. As a new entrant, it is more difficult for us to validate our technology through testing of product."[76] Lack of access to DoD funding may be particularly challenging for early-stage companies that are founded and led by technologists who may not have the requisite business skills to set the business on a path to sustainability through the commercial market.[77]

DoD has funding available to support development, but much of that funding is concentrated in the early stages of technology development. For example, DARPA does "the fundamental research, the proof of principle, and the early stages of technology development that take 'impossible' ideas to the point of 'implausible' and then, surprisingly, 'possible.'"[78] DARPA typically does this through contract vehicles and agreements including Federal Acquisition Regulation (FAR)-based contracts, Broad Agency Announcements (BAAs), SBIRs, OTAs, grants, Cooperative Research and Development Agreements (CRADAs),[79] and prize challenges with industry and universities. DARPA will also partner with the services and other DoD components to conduct technology demonstrations, establish joint program offices to oversee a program, and support transition of early operational capabilities, and DARPA has established programs such as the Embedded Entrepreneur Initiative to support the maturing of the entrepreneur's business processes and exposure to venture capital investors for potential investment.

---

[74] Interview 25, Interviewee 31, September 21, 2021.

[75] Interview 20, Interviewee 24, September 15, 2021.

[76] Interview 20, Interviewee 22, September 15, 2021.

[77] Interview 46, Interviewee 61, November 17, 2021.

[78] DARPA, "Creating Technology Breakthroughs and New Capabilities for National Security," August 2019.

[79] A CRADA is "a legal agreement between a federal laboratory and industry used for the transfer of commercially useful technologies from federal laboratories to the private sector and to make accessible unique technical capabilities and facilities." Scott Ulrey and Diane Sidebottom, "Other Transactions Authority," DARPA Contracts Management Office, presentation, May 2022.

As another example, the SBIR and STTR programs provide federal research funds to domestic small businesses on a competitive basis to support R&D. To participate in these programs, a small business must meet eligibility requirements set out in federal regulations, and several interviewees noted with frustration that SBIR exclusion rules are particularly challenging for start-ups:

> The small business is under 499 people, and you're considered eligible for SBIR, and yet if you are a four-person company with $1 million from a VC, you're not eligible. The whole thing makes no sense whatsoever. We were not choosing; in a start-up you don't have a choice [whether to accept VC funding]. Our options were [either] here's a VC term sheet or wait an unknown time for SBIR funding. So, you take the funding and deal with SBIR ineligibility later. I assume lots of good companies can't work with the government because of that. I'm not sure why that rule exists.[80]

While VC-backed entities are not strictly prohibited from applying to the SBIR/STTR program, there are limitations on the percent of ownership and control that venture capital operating companies can have over the small business.[81] There are exceptions to these rules, but only some parts of DoD accept applications by majority VC-owned entities. DARPA, the Navy, and the Air Force participate in the majority VC ownership authority.[82]

SBIR/STTR funding is capped. As of November 2021, federal agencies may issue a phase I award to "establish the technical merit, feasibility, and commercial potential of the proposed R/R&D efforts" up to a maximum of $275,766, though these awards "are generally $50,000 to $250,000 for six months (SBIR) or one year (STTR)."[83] Agencies may issue a phase II award to continue R/R&D up to a maximum of $1,838,436 (without a waiver), but those awards are generally $750,000 for two years.[84] These funding levels can be helpful to new entrants, but there is no obvious step after a SBIR/STTR phase II award.

DoD may not be getting the full return on its investments in SBIR in part because SBIRs are difficult to transition to programs because of the long lead time for developing budgets. Technology identified as promising through SBIR will often require additional development work, which requires that funds for that effort must be programmed into the budget, which can result in a two- to three-year gap in funding due to the budget cycle. Indeed, a 2019 study by TechLink, commissioned by DoD, found that about 58 percent of DoD's SBIR/STTR phase II contracts between 1995 and 2018 resulted in sales to a final customer; however, only

---

[80] Interview 43, Interviewee 56, November 16, 2021; Interview 41, Interviewee 53, November 10, 2021.

[81] U.S. Small Business Administration, SBIR/STTR America's Seed Fund, "Frequently Asked Questions," webpage, undated a.

[82] U.S. Small Business Administration, SBIR/STTR America's Seed Fund, "VC Ownership Authority," webpage, undated d.

[83] U.S. Small Business Administration, SBIR/STTR America's Seed Fund, "The SBIR and STTR Programs," webpage, undated b.

[84] U.S. Small Business Administration, undated b.

23 percent of these awards represented direct contracts to sell a product or service to a military customer, while 13 percent required additional follow-on development.[85] However, it is important to clearly define the parameters for return on investment in SBIR. According to one metric—commercialization of initial SBIR awards—DoD has achieved a return-on-investment of nearly eight to one. By contrast, DoD's return on investment from transitioning SBIRs to programs or final product sales is considerably more mixed.

One interviewee noted:

> DoD has made $2 billion in SBIR investments—the largest in the government—but we can also leverage investments made by other agencies. A lot of money has been invested in SBIR projects but the platform [into which the technology would be integrated] wasn't ready for the technology . . . they were five years early. So now we have a phase II warehouse of technology targeted for the Joint Strike Fighter that isn't going anywhere. We have a data warehouse of phase II topics. . . . Programs should mine the data better, and they would find a technology that would solve their needs really quickly.[86]

Another government interviewee noted that he often sees the same businesses cycling through SBIR: "You'll get different opinions from government folks on SBIR. Some think it's great; some say 'meh. . . .' You end up seeing the same people [receiving SBIR awards] a lot. Is it innovative if it's the same players?"[87] One early-stage venture founder concurred that insiders have advantages in SBIR applications: "[There is a] disadvantage to outsiders, even though SBIR is supposed to be fair and open competition. [Before the] solicitation [insiders] can talk to the government sponsor, and the government tailors the proposal to that company and technology. Outsiders then apply to the solicitation. But the government . . . already knew who they were going to go give it to."[88] While our research did not independently verify these points, we heard related comments from several interviewees.

From the business's perspective, there may be no obvious next step after SBIR. One small business we spoke with had successfully used a SBIR to conduct a demonstration of a complex maneuver in a volatile environment with clear military applications, but it has yet to find a willing transition partner and is continuing the maturation of its technology using internal R&D funds. This same company was able to get a version of that technology integrated into the planning for a service program of record—their system was selected to provide a specific capability to be incorporated in the future—but funding on a contract does not yet exist.[89] In some cases, these companies have been trying to gain access to DoD for several years, and even longer than a decade, with limited success. It is important to note as well that these

---

[85] TechLink, *National Economic Impacts from the DoD SBIR/STTR Program 1995–2018*, October 3, 2019, p. 27.

[86] Interview 47, Interviewee 63, December 2, 2021.

[87] Interview 14, Interviewee 16, September 2, 2021.

[88] Interview 20, Interviewee 24, September 15, 2021.

[89] Interview 49, Interviewee 66, January 13, 2021; Interview 50, Interviewee 69, January 13, 2021.

examples were identified as "success stories" of technologies by DIOs identified as useful for military applications, not simply companies with a poor value proposition.

We observed that there is limited support available for continued development of technology beyond a prototype, including testing and proof-of-concept demonstrations:

> There's a real thing that starts to happen around TRL-5 or TRL-6, where you need external funding to be taken seriously. That could come from DoD, but there is no big pot of money to move things from TRL 5–7, or you have to create an environment where you will get a dual-use . . . funder at the investor and seed level. The DoD has not solved for either of those [situations]. There is more engagement on the former; there are VC businesses that are willing to make that investment, but mostly at the seed stage. If the department will not solve this [issue] for themselves, most VCs will not invest.[90]

Some innovation organizations are addressing this space. For example, as discussed in Chapter 3, CP/RRTO aims to create faster pathways for concepts to become capabilities by identifying promising emerging technologies and capabilities, supporting prototyping and experimentation to mature them, and transitioning them into programs.[91] DARPA's Small Business Programs Office has a Transition and Commercialization Support Program intended to maximize the potential for companies to move their technologies into other R&D programs for further maturity.[92]

In addition to these organizations, the DoD Rapid Defense Experimentation Reserve (RDER) initiative, announced in 2021, is intended to facilitate prototyping and experimentation to support joint warfighting concepts.[93] In each of these examples, there may still be a two- to three-year gap in funding from the demonstration to getting an approved requirement into the planning, programming, budgeting, and execution (PPBE) process to be funded.

We interviewed the founder of one early-stage venture that solved the early-stage funding problem on its own. They lobbied Congress to get enough money to support testing the product with potential users. They also received significant VC backing to further mature the technology and product. Knowing that user feedback is critical, they "paid for testing [them]selves . . . because we knew we wouldn't get anywhere without it. So, we took it on the road and got soldiers' feedback. That helped us succeed [and] to develop a product that soldiers want and will eventually become a program of record. Soldiers loved it. . . . We went to 15 units and have 20 more units that signed a letter of support. That changed the tune of higher-ups. People are reaching up and saying that's what they want."[94] For most businesses,

---

[90] Interview 23, Interviewee 29, September 20, 2021.

[91] RRTO, 2021.

[92] Defense Advanced Research Projects Agency, "Small Business Programs Office Transition and Commercialization Support Program Fact Sheet," May 21, 2020.

[93] Vergun, November 8, 2021.

[94] Interview 25, Interviewee 31, September 21, 2021.

however, the lack of funding to continue development may mean their company cannot be sustained, or the company refocuses on the commercial market.

In many cases, DIOs are only tangentially supporting the later-stage maturation of a technology. DIOs rely predominantly on facilitating introductions to other stakeholders to support advanced prototyping and testing in realistic operational environments. ARCWERX is limited by its funding (which is exclusively from O&M) in its ability to support prototyping, but like NavalX, it can make connections to service test platforms. A few organizations, like CP/RRTO and AAL, have access to or control over RDT&E funding, which facilitates direct support to prototyping and field testing. We found that other DIOs, however, point to their accelerator programs (e.g., xTech) as a mechanism for further maturation, but accelerators are more often focused on business maturation, improving a company's skills in selling to DoD, and giving them exposure through events like the annual Association of the United States Army (AUSA) conference to a large potential customer base.

Together, these challenges lead to gaps in funding and delays. Technology development may stall, and businesses may struggle to remain viable.

These challenges tie to all development and adoption functions.

Most Existing Opportunities Are Overly Burdensome to Pursue

Once those funding opportunities are identified, most existing DoD opportunities are difficult and disproportionately costly for new entrants to pursue, which deters them from working with DoD. Burdensome processes are disproportionately costly to small new entrants, which often do not have dedicated teams (unlike traditional defense suppliers) to develop proposals:

> The hesitancy is, on the other hand, if you want to tap into some companies that just look at DoD as a monolith and don't want to compete with Northrop Grumman and Lockheed Martin. They're not even going to bother since they don't have the luxury of time and money generating technical papers. Everything is hours, which means dollars.[95]

Many small technology companies find the DoD environment, with its insular marketplace and technical and procedural barriers, unappealing and impenetrable.[96] These barriers may include cost accounting standards, budget processes, contract reporting, FAR- and Defense Federal Acquisition Regulation (DFAR)-required contract clauses, intellectual property, and technical data rules.[97] Given these bureaucratic hurdles, some note that the DoD

---

[95] Interview 31, Interviewee 41, October 6, 2021.

[96] Interview 20, Interviewees 22 and 23, September 15, 2021; Interview 27, Interviewee 34, September 24, 2021; Interview 31, Interviewee 41, October 6, 2021; Interview 46, Interviewee 61, October 19, 2021.

[97] Giles K. Smith, Jeffrey A. Drezner, William C. Martel, James J. Milanese, W. E. Mooz, and E. C. River, *A Preliminary Perspective on Regulatory Activities and Effects in Weapons Acquisition*, Santa Monica, Calif.: RAND Corporation, R-3578-ACQ, 1988; Jeffrey A. Drezner, Irv Blickstein, Raj Raman, Megan McKernan,

market is not worth the effort to pursue: "Working with DoD is seen as too long-lead, uncertain, small profit. DoD misses out because they aren't seeing these tech companies that won't even bother with DoD as a market."[98] Another noted: "Many from industry or the cyber community don't want anything to do with the government since they've heard horror stories."[99]

Those new-entrant companies that are willing to apply for DoD opportunities often do not understand what the government is seeking or how to write proposals, and therefore they are not competitive with established sellers.[100] Compounding these challenges, funding decision processes are slow, even when they are designed to be faster than standard DoD processes. The promise of 30 days to decision and 60 days to contract, for example, is more likely to go unfulfilled. Compared with a "shark tank" (a gathering of venture capitalists, each of whom is deciding whether to invest in a particular firm) where a funding decision can be made the same day as the pitch, the defense innovation ecosystem cannot compete.

These challenges tie to all development and adoption functions.

Many Factors (including Funding, Misalignment with Budget Process, Technical Failure, and Lack of a Formal Requirement or Customer Who Is Ready) Contribute to the "Valley of Death"

"Transition is aspirational. It's not really happening."[101]

The "valley of death" remains the most significant impediment in the CTP. Businesses trying to sell to DoD must cross the "valley of death," a term of art typically defined as a gap, pause, or delay in development, production, or fielding when a technology has been demonstrated at TRLs 5 or higher and is technically ready to be incorporated into a system design or program and transitioned from development to production, fielding, operations, maintenance, or support activities. The problem occurs due to a misalignment between the timing, resourcing, and formality (documentation and reporting needs) of early-stage development activities and the timing, resourcing, and formality of processes and decisions associated with programs of record.

The valley of death phenomenon presents challenges and costs to both DoD and commercial firms. The delay interrupts the flow of development and test activities, slowing progress

---

Monica Hertzman, Melissa A. Bradley, Dikla Gavrieli, and Brent Eastwood , *Measuring the Statutory and Regulatory Constraints on Department of Defense Acquisition: An Empirical Analysis*, Santa Monica, Calif.: RAND Corporation, MG-569-OSD, 2007; Section 809 Panel, *Report of the Advisory Panel on Streamlining and Codifying Acquisition Regulations*, Vol. 1, January 2018, Vol. 2, June 2018, Vol. 3, January 2019.

[98] Interview 46, Interviewee 61, October 19, 2021.

[99] Interview 14, Interviewee 16, September 2, 2021.

[100] Interview 13, Interviewee 15, August 27, 2021.

[101] Interview 5, Interviewee 6, August 5, 2021.

and creating inefficiencies, and it may adversely affect the financial viability of a firm or result in termination of the technology development. The valley of death may also result in a delay in fielding potentially useful capabilities and may result in cost or schedule increases to an existing program of record trying to integrate the technology as well as other related or interdependent programs.

There are several situations of particular interest here. First, the valley of death may prevent the transition of a technology with identified military utility and a program of record wanting to incorporate it. This direct impact is most likely if the technology development is terminated or the firm goes out of business, Second, the valley of death may prevent a technology with military utility from gaining the exposure it needs to demonstrate that utility. In the latter case, DoD may not yet know that it needs the capability the innovative technology provides. Finally, the technology may not transition simply because a military department or operational user has other priorities in a constrained budget environment. Only the first and second situations are problems from the DoD perspective; the third is not a problem because there was an explicit decision to not transition the technology at this time.

In the CTP framework, we have defined three ways businesses can transition their technologies to the department for operational use: (1) fielding at scale, as part of a post–milestone C production program; (2) limited fielding, as part of a rapid acquisition program; and (3) fielding as a residual capability (i.e., prototypes) resulting from a prototype demonstration, an advanced concept technology demonstration, or from the Middle Tier Acquisition (MTA) pathway. From a business's perspective, navigating the valley of death is often lengthy—one or more years, not months—and often unsuccessful.[102] Moreover, prototyping and demonstration becomes more expensive, and the burden of proof for operational viability increases in the TRL 5–7 range: "That's when dollar values go up, the proof gets more expensive as you climb the TRL scale. An engine is not fast or cheap to develop. The research labs are set up for low TRL, same as [the DIOs], and all acquisition groups are TRL 7+. So, there's a very significant gap in this TRL 5–7 range."[103]

The valley of death is often described as primarily a funding problem. For example, representatives from one commercial business with experience in DIO programs described their experience as a technology that was "not a real program . . . [but] under these seedlings [programs] that didn't have resources for transitioning anything." Following the completion of their SBIR contract with a DIO, "[the DIO] just dropped us off at end of the third phase."[104] However, the notion that funding alone drives the valley of death is a misconception, though it is an important factor; new technologies must compete in the budget process with other

---

[102] James M. Landreth, "Through DoD's Valley of Death," Defense Acquisition University, February 1, 2022.

[103] Interview 43, Interviewee 56, November 16, 2021.

[104] Interview 43, Interviewee 56, November 16, 2021.

new and established technologies. Technologies may fail to cross the valley of death for many reasons, including lack of budget (the service or user does not have the money to purchase the technology), misalignment with the budget process, technical failure (the technology fails to reach the appropriate maturity level), no formal requirement for the capability represented by the technology, inability to identify a matching requirement, inability to find an end user who needs the technology, no direct applicability to a specific program manager's capability portfolio, a poor record of demonstration or testing in an operational setting, or a prime contractor's unwillingness to accept integration risk on an existing program of record.[105] Moreover, it is challenging for DIOs to match a business to a customer that has identified a specific problem, has an allocated set of funds available, and has realistic expectations about what they can get out of the program. According to one interviewee, "The number-one issue is finding a customer who is ready."[106]

The variety of transition pathways and possible obstacles to transition led one interviewee to note that there are "multiple valleys of death" that must be defined and navigated.[107] Nevertheless, it is widely acknowledged that technology emerging from DIOs face similar transition challenges, irrespective of their different missions and technologies: "A lot of innovation organizations, NavalX, AFWERX, Army Futures, are all doing different stuff. But they recognize the same problem. I don't know if there is a good solution."[108]

Another common misconception is that the valley of death persists because contracting authorities are insufficient. For example, one interviewee noted that the "only way to get to a program of record is to jump a number of stages. You can use OTA [Other Transaction Authority] as a bridge, but you sometimes have to become a subcontractor to a prime. If you want to get to a program of record status, it's almost impossible. The closest you can get is to be a good subcontractor."[109] The bulk of our research, however, indicates that this is not quite right: Existing contracting authorities, including nontraditional pathways such as OTA and MTA, have been effectively leveraged to bridge the gap between TRLs 3-6 and TRLs 7-9 and are common mechanisms that acquisition stakeholders employ to provide follow-on development work to businesses that successfully complete DIO accelerator and challenge programs. However, existing contracting authorities are nevertheless being underutilized to bring commercial technologies into the department.

Efforts to address the valley of death may fail because they focus on only one of these factors, such as funding.

These challenges tie to adoption functions: transition to fielding.

---

[105] John Dillard and Steve Stark, "Understanding Acquisition: The Valley of Death," United States Army Acquisition Support Center, October 6, 2021.

[106] Interview 12, Interviewee 14, August 26, 2021.

[107] Interview 19, Interviewee 21, September 15, 2021.

[108] Interview 19, Interviewee 21, September 15, 2021.

[109] Interview 3, Interviewee 3, July 30, 2021.

The Valley of Death Persists in Part Because No DoD Organization Has Visibility into All CTP Activities or Responsibility for CTP Outcomes, and Gaps Between Stakeholders Hinder Adoption

We found that key reasons that the valley of death persists are that (1) no single organization has visibility into CTP activities enterprisewide or responsibility for CTP outcomes; and (2) gaps between innovation organizations, requirements and capability developers, end users, PMs and PEOs, procurement and decision authorities, and existing incentives for these stakeholders contribute to the valley of death and hinder adoption.

Promising technology that has demonstrated practical application often founders because the links between the organization sponsoring the work (e.g., a service lab) and the potential customer and program office are weak. One company noted that "we built [the technology] . . . and when they went back into the Navy laboratory system, it became pretty clear. [The technology wasn't going] to come out of the laboratory system and . . . became Hangar Queens, not because of technical reasons, but just because there was no transition mechanism."[110] As one interviewee observed, for some technologies, "there's not someone responsible for it within DoD. No one program office is responsible for that. . . . That leads to a chicken-and-egg question. It's not my responsibility, so whose is it?"[111]

DIOs are not able to solve these problems independent of other CTP stakeholders. For example, the problem curation and solution time frame can hinder transition, particularly given the long lead time for innovation: "Transition is an ongoing challenge because we don't curate with the services; we look to see what they need in ten to 15 years, but they might not recognize they need it yet."[112] Likewise, multiple representatives from commercial businesses that have participated in various DIO programs indicated that some DIOs underinvest in matching functions. Additionally, many DIOs prioritize early-stage concept development and maturation over follow-on development, as noted above, which leads them to view businesses that have completed initial SBIR phase I/II contracts as "successes" even if the technology has yet to progress into later-stage development. When commercial businesses are neither matched with a potential DoD sponsor nor offered follow-on contracts by the DIO, they are left with few options to continue development. As a representative of one new-entrant venture noted:

> That is the whole crux of the problem. Defense innovation organizations don't help [companies enter and navigate DoD all the way through to a contract]. That's our story. In our case, luck and a few connections made a big difference. The big problem is that you're putting tens of thousands of dollars into early concept work, but it doesn't really help mature the concept or technology. You might be able to find some mentorship. . . . Defense innovation organizations helped out a little bit at the beginning [as we were seeking early funding, helping to educate us about DoD processes], but it's harder once you have to get

---

[110] Interview 49, Interviewee 67, January 13, 2022.

[111] Interview 11, Interviewee 12, August 25, 2021.

[112] Interview 14, Interviewee 16, September 2, 2021.

contracts. The harder part, [the defense innovation organizations] get less involved. They don't tell us we can go talk to units. I think they can do much better. We don't need hand-holding, but DoD is a big scary beast. They're like, "Go figure it out."[113]

Another representative from a commercial business noted, "There is no clear path to a contract. You have a need, and we build it, but then it's crickets. They don't teach you how to get onto a program of record. We're on the precipice of cracking the shell. It would make sense when you put all this money, you want them to succeed."[114]

Other interviewees disputed the notion that DIOs are not well equipped to transition commercial technologies into the traditional acquisition system. One interviewee acknowledged that the valley of death is a genuine problem that "has a way to go," while still contending that the DIOs "are on the outside, to help guide and advise industry on how to figure that out at the same time. We made a lot of progress over the past year."[115] To this end, some DIOs have established transition cells to effectively guide transitioning technologies into the acquisition pipeline.

Although many innovation organizations talk about transition planning and think about potential routes for DoD adoption from the very start of a business's engagement with the DIO as part of their work,[116] most DIOs do not bring in the traditional acquisition and requirements communities "as key stakeholders from the beginning" as they identify technologies of interest and conduct early-stage CTP functions.[117]

DIOs lack consistent buy-in from the traditional communities in large because incentive structures are misaligned between DIOs and the traditional acquisition and requirements stakeholders. We observed that in general, these stakeholders are inclined against incorporating early-stage technologies in their programs. As one interviewee observed, "If you don't have buy-in from service acquisition folks, it will not transition. That is a challenge."[118]

In particular, program managers are not incentivized to support development or fielding of commercial technologies. PMs would seem to be the best avenue to support further development of promising technologies because they have the right type of funding to move past technology demonstration (e.g., BA 6.4 or 6.5), usually have an approved requirement or validated need, and a program generally formulates a plan to mature the technologies being used in the system. However, PMs can only spend their budget on an approved program, with very little ability to shift money between accounts (e.g., RDT&E is distinct from procurement funds; BA 6.3 advanced technology development is different from 6.4 advanced component

---

[113] Interview 25, Interviewee 32, September 21, 2021.

[114] Interview 25, Interviewee 31, September 21, 2021.

[115] Interview 18, Interviewee 20, September 15, 2021.

[116] Interview 2, Interviewee, 2, July 30, 2021.

[117] Interview 28, Interviewee 35, September 28, 2021.

[118] Interview 28, Interviewee 35, September 28, 2021.

development and prototyping).[119] For program managers to support development of a technology, they would typically need that money to be programmed into the budget, which occurs two years in advance under the normal cycle. Generally, a program of record cannot start without an approved requirement in some form, which may add another year, and technology development and maturation would need to have occurred—that is, the technology would need to be demonstrated—or the next level of budget authority would have to have been programmed into the budget.

Indeed, we found that program managers are often unwilling to support development of technology unless there is a direct alignment to their program, and even then, they may do so reluctantly due to the changes it may cause to program schedule and risk. Technology selection for system design and development is usually accomplished early in a program, and it is often the prime contractor that makes that selection because the risk associated with integrating a technology (i.e., a subsystem) is borne by the prime contractor. Program managers are therefore reluctant to use other contracting authorities and direct a vendor to incorporate a new, innovative technology (especially in later stages of system development) or use a particular subcontractor in an ongoing program of record, which can introduce integration risk to the program and transfer risk from the prime contractor to the government (the PM). Indeed, a significant impediment to transition remains the "huge amount of risk involved, where [the innovation] has to jump track at the end of the process to take the thing that is innovative and get it back into [DoD] to repopulate."[120] This means that in most cases, for a technology to be incorporated into a program of record, the service requirements and acquisition community and potential prime contractors must be engaged by DIOs and early-stage venture businesses very early in the process, likely during the identification phase.

Some DIO representatives specifically cited the lack of buy-in and misaligned incentives from stakeholders in the traditional communities as key the impediments to transition: "I have seen no evidence there is a [technology] supply-side problem and no evidence it's a contacts or network engagement problem. It's once we get people in the door [in DoD]."[121] Taken together, these challenges mean that technologies that do not have a clear and natural fit with one program or portfolio but would benefit DoD overall have a hard time finding a champion.[122] Even businesses that have successfully received DoD support to develop their technologies are left without options to field. We found that the result of both the disconnect between stakeholders and the misaligned incentives is that fielding ("adoption") functions are gapped almost entirely. As a result, technologies stall in the valley of death without a clear path to continued development or transition.

These challenges tie to adoption functions: transition to fielding.

---

[119] Defense Acquisition University, "Research Development Test and Evaluation (RDT&E) Funds," webpage, undated b.

[120] Interview 23, Interviewee 29, September 20, 2021.

[121] Interview 23, Interviewee 29, September 20, 2021.

[122] Interview 46, Interviewee 61, November 17, 2021.

Challenges: The Bottom Line

Defense innovation organizations are doing many things well. These organizations are searching far and wide and identifying interesting technologies with potential military applications. Innovation organizations are helping technology innovation hubs like Silicon Valley so that new-entrant businesses become aware of DoD. Innovation organizations are conducting market research and outreach events to develop their innovation networks and identify promising technology. They facilitate the identification and refinement of defense problems to make them more accessible and understandable to an audience less familiar with DoD. DIOs have also developed multiple approaches to help with solution concept generation.

But innovation organizations cannot independently overcome the valley of death, as they do not control requirements, programs, or budgets. Because the valley of death persists, these organizations continue to face challenges in connecting maturing technologies with other defense stakeholders who can continue that maturation (though some have had some success) and even more difficulty in securing a smooth transition between technology development and uptake into programs. This is not for lack trying, but the inherent long lead times for programming and budgeting and a perception of risk aversion on the part of program managers has led many DIOs to define their transition activities as a handoff to someone else to carry the technology forward, whether through additional prototyping or technology maturation, but rarely as a direct transition to a program of record. One innovation organization representative put it candidly: "I can't speak to transition because it has been aspirational."[123]

The biggest gap in the CTP is that no one has both responsibility and authority for solving the valley of death and transitioning innovative commercial technology at scale. No single office or stakeholder owns the entire problem. There is no contract vehicle, acquisition activity, or funding specifically targeted at the valley of death problem. Incentives for the relevant CTP stakeholders (e.g., requirements developers, PMs, PEOs, and end users) are not in place to overcome these gaps. There are currently no formal mechanisms or requirements for information sharing, coordination, or collaboration across the defense innovation ecosystem or CTP. Rather, information sharing, coordination, collaboration, and feedback mechanisms are ad hoc and often based on personal relationships.

This implies that the DIE and CTP are not functioning ecosystems, despite observers' frequent use of the term "innovation ecosystem." Ecosystem processes are nonlinear; there are multiple feedback loops and node interactions that drive how well the ecosystem works. Understanding an ecosystem requires a holistic and dynamic (adaptive) framework. Ecosystems are also self-regulating, usually through some form of information sharing. While these conditions are not present in the CTP or DIE today, by fostering some of these feedback loops and relationships, DoD may improve CTP outcomes.

---

[123] Interview 5, Interviewee 6, August 5, 2021.

CHAPTER 5

Approaches to Strengthen the Commercial Technology Pipeline

Drawing on the CTP model and analysis of challenges, the research team developed and evaluated alternative approaches to mitigate identified challenges and strengthen the DoD CTP. We used a policy game construct (described in Appendix C) as well as our analysis of the literature review and interviews to evaluate a select number of alternative approaches or policy levers. We were able to test and evaluate many of the policy levers directly in the game; even when we could not, the game often provided some insight into the effect of other policy levers. Similarly, analysis of literature review results and interviews provided some direct insight into policy levers that we could not test directly in the game and lent additional support to findings from the game.

This chapter presents the alternatives we developed and tested and our findings relating to each. It is divided into two sections: alternatives relating to (1) cultivating desired characteristics of a well-functioning CTP; and (2) strengthening CTP identification, development, and adoption functions. Related recommendations appear in Chapter 6.

Identifying Alternative Approaches

We sought to identify alternative approaches to strengthen the CTP that were feasible and plausible to implement. Toward those ends, we assumed that the basic contours of the CTP would remain the same and did not consider alternatives that would require a radical reset of the current landscape of innovation organizations; CTP stakeholders; or requirements, acquisition, and budgeting processes.

With these framing considerations in mind, we drew on all data sources, including the literature review, interviews, the DIO deep dives, and technology case studies, to identify alternatives for consideration. In each of these sources, we identified suggestions to improve one or more CTP functions or characteristics. We analyzed the challenges that we previously identified, mapped alternatives to those challenges, and augmented this list by developing additional alternatives to address challenges.

The list of alternative approaches that we considered included changes to the missions, authorities, processes, organizational structures, human capital (both leadership and workforce), collaborative relationships, and resources of existing defense innovation organizations; creation of or elimination of innovation organizations; and changes to the processes and structures through which defense innovation activities are overseen.

Tables 5.1 and 5.2 list the approaches evaluated. Each option is described in detail later in the chapter.

Evaluating Approaches Through a Policy Game

To select and develop the policy levers to test in the game, we drew from the longer list of alternative approaches that we had developed and down-selected to identify potential policy levers that we assessed would strengthen the CTP, that the OSD has authority to implement (only the option of a flexible funding account would require congressional approval), that we deemed plausible, and that could be represented in the game. The two packages of policy levers that we tested on day 1 and day 2 sought to foster the characteristics of a well-functioning CTP and address some key challenges that hinder adoption of commercial technology:

- The game construct gave all stakeholders a shared mission of creating more commercial technology opportunities and pushing promising technology all the way through the CTP.
- We issued policy guidance that specified goals, objectives, and desired outcomes for the CTP; defined stakeholder roles and responsibilities; and specified metrics and accountability mechanisms.
- Players shared information throughout the game (e.g., promising technologies, available resources and programs, priorities and focus areas, collaboration opportunities).
- Players were incentivized to collaborate to realize CTP goals, objectives, and outcomes, and we created feedback mechanisms.

The game revealed insights about the status quo, challenges, and policy levers through which OSD could strengthen the CTP.

Approaches to Cultivate Desired CTP Characteristics

We found that the desired characteristics of a well-functioning CTP are not present and that most CTP enabling functions—including policies, guidance, planning, funding and investment, coordination, navigation support, information sharing and reporting, data collection and analysis, oversight, and provision of hard infrastructure to support development—are either not occurring regularly or are occurring on an ad hoc basis. DIOs tend to be operational agencies focused on executing their core CTP functions. Some DIOs also provide targeted funding and navigation services. However, responsibility for enabling functions—setting policy and guidance; allocating funding; providing mechanisms for information sharing, coordination, and collaboration across the DoD enterprise; ensuring accountability—lie at a higher organizational level within both OSD/Joint Staff and the services.

TABLE 5.1

Approaches Evaluated to Cultivate Desired CTP Characteristics

Option	Related CTP Core and Enabling Functions
Foster a shared sense of mission across all CTP stakeholders	Enabling: policies and guidance, planning and plans, oversight
Develop and promulgate strategy, plans, policies, and guidance for the CTP to establish common goals, objectives, and outcomes	Enabling: policies and guidance, planning and plans, oversight
Define and communicate roles and responsibilities for the CTP, ensuring there are no gaps in key functions	Enabling: overall functioning of the CTP and each of the core and enabling functions
Facilitate information sharing, coordination, and collaboration across CTP stakeholders	All CTP functions
Implement incentive structures, including metrics and accountability mechanisms, to align CTP stakeholders to CTP goals, objectives, and desired outcomes	Enabling: funding and investment, data collection and analysis, oversight

We recommend that DoD cultivate desired CTP characteristics, but we note that it will be important to balance any additional centralized guidance and coordination with fostering the agile environment that technology innovation requires. Table 5.1 lists the approaches we evaluated to foster desired CTP characteristics. Each option is described in detail in the sections that follow.

Foster a Shared Sense of Mission Across All CTP Stakeholders

While innovation organizations view themselves as playing a role in the identification, development, and adoption of commercial technologies, they have disparate missions and do not view themselves as part of an integrated pipeline or system. Key stakeholders in the traditional acquisition, requirements, budgeting, and end-user communities do not see themselves as part of the CTP at all and are not working with innovation organizations as these organizations identify technologies of interest and conduct early-stage CTP functions. Box 5.1 describes an approach we evaluated to address this issue.

BOX 5.1

Foster a Shared Sense of Mission Across All CTP Stakeholders

Hypothesis: By fostering a shared mission, DoD can lay the foundation for information sharing, coordination, and collaboration that is required for the CTP to function.

How the policy lever was modeled: We communicated clearly to game participants that their mission was to try to identify, develop, and field promising technologies, pushing them all the way through the pipeline. The game construct reinforced this sense of shared mission: We provided all participants information on a delimited set of technologies and "chips" representing funding that each participant could invest. We built a game board that represented CTP core functions and brought participants together virtually for two four-hour blocks to plot their investments (one investment per game "move"). The research team ran a "white cell" that adjudicated the outcomes of each move, and we displayed the outcomes on slides viewable by all participants so that each could track the movement of technologies through the pipeline.

Links to CTP functions: This approach links to CTP enabling functions: policies and guidance, planning and plans, and oversight.

Results

To provide context for the results, it is important to note that the game construct radically simplified reality and could not be mimicked in the real world. There are hundreds of CTP stakeholder organizations and an ever-growing pool of potentially relevant technologies to consider, and it would not be possible to bring all stakeholders together for a live session to discuss, deconflict, and coordinate investment decisions to maximize pipeline throughput.

The game findings nonetheless suggest that DoD could strengthen the CTP by fostering a sense of shared mission among CTP stakeholders. Participants representing innovation organizations, program managers, and end users all responded positively to the shared mission of creating collaborative opportunities and pushing promising technologies all the way through. To varying degrees, all players sought to move technology through the pipeline. PMs were the least forward-leaning, a point we revisit in connection with incentives below.

Develop and Promulgate Strategy, Plans, Policies, and Guidance for the CTP to Establish Common Goals, Objectives, and Outcomes

At present, there is no DoD-wide strategy for innovation, innovation priorities, innovation plans or planning function, or guidance. Nor has DoD promulgated strategic goals and outcomes for the CTP to which stakeholders can align their plans and activities. As a result, DIOs pursue disparate problems and technologies that may not be the highest priority or most impactful. Box 5.2 describes an approach we evaluated to address this issue.

BOX 5.2

Develop and Promulgate Strategy, Plans, Policies, and Guidance for the CTP to Establish Common Goals, Objectives, and Outcomes

Hypothesis: If OSD sets strategy and provides guidance, innovation organizations and other CTP stakeholders will pursue problems and technologies that are deemed to be of the highest priority by leadership. They will share information and collaborate more readily, and DoD will be able to oversee and hold stakeholders accountable, using the same reference points and understandings of key concepts such as transition. This would result in better outcomes from the perspective of the pipeline as a whole.

How the policy lever was modeled: We modeled this policy lever in the acquisition policy game by providing participants with a mock-up policy memorandum outlining broad strategy and guidance relating to DoD innovation. On the first day of the game, we did not enforce compliance with this guidance. On the second day, we provided a mock-up DoD Directive 5000.97, "DoD Commercial Technology Pipeline," specifying goals, objectives, technology priorities, a timeline, metrics, milestones for all CTP stakeholders, and an oversight structure. We signaled that compliance was mandatory, and during the game the research team "white cell" lightly enforced compliance with stated technology priorities by tying compliance to a flexible funding pool (discussed below).

Links to CTP functions: This approach links to CTP enabling functions: policies and guidance, planning and plans, and oversight.

Results

All players heeded the overall game objective of moving technologies through the pipeline. The mock OSD priorities were a consideration for players in making some investment decisions, but players demonstrated a willingness to depart from OSD priorities. In relation to the policy guidance provided during the game, participants had the following comments:[1]

- "What things are in alignment with new notional deputy's priority areas?"
- "I recognize we have the 'shall' here, but [my organization] will probably go rogue against priorities about 50 percent [of the time]."
- "I didn't really look at the [OSD] guidance that much."
- "To be honest, I didn't even think about the [OSD priorities]. I had no consideration for that as I was going through."

---

[1] RAND NDRI acquisition policy game, March 9–10, 2022.

Discussing their real-world behavior, participants noted the following:[2]

- "We don't necessarily follow R&E [Office of the Under Secretary of Defense, Research and Engineering] priorities. We try to find things people haven't done before."
- "Policy guidance does matter in terms of incentives. My office [has] flexibility similar to what [another innovation organization representative] said of not following the guidelines exactly or interpreting them willfully. But a few times, [we were] looking at which technologies line up against programs or policy memos. But I at least had some flexibility around that and will execute that. It's a factor. I don't think leadership wants us to blindly execute things; they want us to be thoughtful. There's a balance there."

The impact of the guidance in shaping behavior appears to have been limited by participants' competing priorities and the white cell's light-touch approach to enforcement. Program managers were least willing to change behavior based on OSD guidance, without other changes to their existing incentive structures. **We concluded that DoD guidance can reinforce a sense of shared mission and common objectives, but the level of compliance with guidance may depend on OSD oversight and enforcement.**

Define and Communicate Roles and Responsibilities for the CTP

CTP stakeholders' roles, responsibilities, areas of focus, and relationships with one another are not well understood. Many traditional acquisition stakeholders such as requirements professionals and program managers do not view identification, development, and adoption of dual-use technologies as a distinct part of their job. This contributes to overlap, duplication, and gaps in CTP functions and also to the valley of death: the failure of many promising technologies to transition to fielding. Box 5.3 describes an approach we evaluated to address this issue.

BOX 5.3

Define and Communicate Roles and Responsibilities for the CTP, Ensuring There Are No Gaps in Key Functions

Hypothesis: By defining and communicating roles and responsibilities for the CTP, DoD can help to ensure that CTP stakeholders understand their roles and responsibilities, understand other organizations' roles in the CTP, minimize duplication and overlap, and ensure that all key functions are performed, minimizing gaps. Increasing information sharing, coordination, and opportunities for collaboration could also reduce duplication by making a solution concept that one stakeholder is supporting visible and available to all CTP stakeholders.

---

[2] RAND NDRI acquisition policy game, March 9–10, 2022.

How the policy lever was modeled: On day 1 of the policy game, we provided participants with a mock-up policy memorandum outlining general roles, responsibilities, and focus areas for defense innovation organizations. The memo also specified roles and responsibilities of other CTP stakeholders at a high level. The research team white cell did not enforce those roles or responsibilities. On day 2, we provided participants with a mock-up DoD Directive 5000.97 that specified the roles, responsibilities, relationships, CTP functions, problem sets, customers, time frame, and regions of focus of CTP stakeholders, including innovation organizations, program managers, and end users. We signaled that compliance was mandatory but did not strictly enforce the guidance.

Links to CTP functions: This approach links to the overall functioning of the CTP and each of the core and enabling functions.

Results

We found that the policy memorandum was irrelevant. Players demonstrated a strong desire to "stay in their lane" and enforced their own roles and responsibilities without reference to the policy memo. However, there was some indication that when players better understood the roles of other CTP stakeholders, they identified more opportunities for collaboration. Thus, we saw again that, **without enforcement, policy guidance appears to have limited utility as a mechanism to shift behavior.** Nonetheless, we assess that as part of a package of policy changes, and when coupled with appropriate enforcement, this type of guidance could strengthen the CTP.

Facilitate Information Sharing, Coordination, and Collaboration Across CTP Stakeholders

The major constraint on technology throughput and adoption (fielding) remains the disconnect between stakeholders. We found that information sharing, coordination, and collaboration across the CTP are ad hoc and often based on personal relationships, resulting in duplication, inefficiencies, lack of awareness of what other CTP stakeholders are doing, and lack of handoff between CTP stakeholders. There is no mechanism to institutionalize the knowledge and experience of individual stakeholders and improve the situational awareness of all stakeholder organizations. CTP stakeholder organizations are in many cases not capturing the value of information that their individual employees have, and DoD as a whole is not capturing the value of information held by CTP stakeholders. When innovation ecosystem knowledge and contacts are personalized, the system risks losing that knowledge when officials move on. Box 5.4 describes an approach we evaluated to address this issue.

BOX 5.4
Facilitate Information Sharing, Coordination, and Collaboration Across CTP Stakeholders

Hypothesis: If DoD facilitates information sharing, coordination, and collaboration, CTP stakeholders will work together more efficiently to move technologies through the pipeline to fielding.

How the policy lever was modeled: We modeled this policy lever in the acquisition policy game on day 1 by allowing game participants to share information as part of decision-making. We required players serving in innovation organization roles to explain the technologies in which they were interested and allowed traditional acquisition stakeholders (end users and PMs) to express interest in technologies also. Those traditional stakeholders also had additional information on requirements and capability gaps that they were permitted, though not required, to share. The research team white cell provided limited guidance, and there was no requirement or designated time for players to coordinate or deconflict their investments. On day 2, we required and provided time for participants to share and discuss their investment choices before those choices were final, and we encouraged participants to coordinate and deconflict their choices. This meant that traditional acquisition stakeholders were more engaged in the innovation organization's early-stage decisions.

Links to CTP functions: Coordination and information sharing are enabling functions that underpin all other CTP functions.

Results

We found that players were eager to collaborate. Program managers were especially eager to get signals from end users that could help them decide whether to move funds from existing programs and which technologies would be of interest. Players expressed interest in coordinating their investments, but without a designated period or mechanism for coordination on day 1, all coordination offers went unmet.

Collaboration was successful in the game under particular conditions. Participants shared a common goal of moving technologies through the pipeline. According to one player, most important is "the objective—you need to emphasize this, and make sure everyone understands the goal and their role in it." Relatedly, we provided participants with a finite number of technologies, and those technologies were common across all players. We designed the game so that collaboration was required to succeed. Players could not get technologies all the way through the pipeline without collaboration, because no individual organization had enough funding, and some participants could not perform certain CTP functions.

We provided a forum for collaboration, and incentivized collaboration by providing additional matching funds to collaborating organizations. Under these conditions, we found that players were eager to collaborate. We observed that while collaboration was messy and time-consuming, it was successful: technologies moved through the pipeline.

Implement Incentive Structures, Including Metrics and Accountability Mechanisms, to Align CTP Stakeholders to CTP Goals, Objectives, and Desired Outcomes

Innovation organization oversight is currently conducted by Congress and innovation organization parent organizations, and no one appears to hold other CTP stakeholders accountable for how well their activities support the identification, development, and adoption of dual-use technologies for military use. The department has not specified metrics for CTP stakeholders to provide insight into how well CTP-related processes are working. Innovation organizations appear to develop and report their own metrics and, in some cases, selectively report favorable information to sustain support and funding. Box 5.5 describes an approach we evaluated to address this issue.

BOX 5.5

Implement Incentive Structures, Including Metrics and Accountability Mechanisms, to Align CTP Stakeholders to CTP Goals, Objectives, and Desired Outcomes

Hypothesis: By implementing incentive structures, including metrics and accountability mechanisms, together with the guidance described above and clarification of roles and responsibilities described below, DoD can incentivize innovation organizations and other CTP stakeholders to work toward CTP outcomes.

How the policy lever was modeled: On day 1 of the game, we specified "easy" output-focused metrics. We required CTP stakeholders to self-report against those metrics and noted that an existing OSD/R&E office would be reviewing those submissions. We found that under these circumstances, the metrics and annual oversight were irrelevant to participants: no one changed their behavior. On day 2, the mock-up DoD Directive 5000.97 described a fictional newly created Deputy Assistant Secretary of Defense for Innovation—DASD(Innovation)—that would develop directive-type policy, strategy, and metrics and oversee CTP functions and stakeholders, monitoring compliance with OSD/R&E technology priorities, collecting and analyzing CTP output and outcome data, and managing a shared database of CTP-relevant information. The directive specified outcome-focused metrics and data reporting requirements for CTP stakeholders and noted that DASD(Innovation) would monitor compliance. Finally, the directive described a flexible funding pool with multiyear fungible funds that participants could apply for to support demonstration and handoff or to maximize transition opportunities. DASD(Innovation) had authority to allocate the flexible funding to participants, with a focus on priority technologies and maximizing adoption opportunities.

Links to CTP functions: This approach links to CTP enabling functions: funding and investment, data collection and analysis, and oversight.

Results

Participants' adherence to guidance differed depending on the level of enforcement and opportunities for funding. With loose enforcement, we found that OSD-specified priorities and metrics informed player investment choices, but players were willing to depart from those priorities; without enforcement, the directive's statement of priorities and guidance became a request. When coupled with the opportunity to apply for flexible funding, however, participants were careful to heed the guidance. We found that participants were eager to access the additional OSD funds, collaborated and strategized to do so, and were careful to meet the requirements for access to those funds, including alignment with OSD strategic priorities. We found that those funds helped players move technologies through the pipeline and resulted in more transitions.

Our research suggests that carrots (such as funding) may be more effective than sticks (such as top-down strategic guidance and oversight) to change incentives for innovation organizations. We found that shifting incentives for program managers may be more challenging, especially if they are limited to funding associated with current programs. Our interviewees and game participants suggested program managers are generally unwilling to incur risk to their programs by pulling money away from them without end-user service leadership. A full examination of possible incentives and alignment of CTP stakeholder interests was beyond the scope of this study, but these topics are worthy of additional study.

Approaches to Strengthen CTP Identification, Development, and Adoption Functions

As noted in Chapter 4, we found that many individual CTP functions are not functioning optimally and that there are both overlaps and gaps in the CTP adversely affecting outcomes. Table 5.2 lists the approaches we evaluated to strengthen CTP identification, development, and adoption functions. Each option and our evaluation of its effect is described in detail in the following sections.

TABLE 5.2

Approaches Evaluated to Strengthen CTP Identification, Development, and Adoption Functions

Option	Related CTP Core and Enabling Functions
Develop more rigorous, comprehensive, and shared approaches to technology scouting	Core: technology scouting Enabling: navigation and relationships
Encourage CTP stakeholders to share problems across the enterprise	Core: identification functions (problem identification, validation, curation; solution concept generation; customer discovery; matching problems to solutions) Enabling function: navigation and relationships
Establish a single, comprehensive, integrated, searchable portal for new entrants to access DoD opportunities	Core: technology scouting and customer discovery, solution concept generation, network development Enabling: navigation and relationships

Table 5.2—Continued

Option	Related CTP Core and Enabling Functions
Establish navigation support services to help new entrants understand and navigate DoD	Core: customer discovery, solution concept generation, all development and adoption functions Enabling: navigation and relationships
Assign responsibility for oversight of the CTP, including transition outcomes, to an organization and give it the authority and budget to execute those responsibilities	All CTP functions
Provide flexible funding to incentivize CTP stakeholder collaboration and to close transition gaps	Core: all development and adoption functions Enabling: funding and investment, coordination, plans and planning

Develop More Rigorous, Comprehensive, and Shared Approaches to Technology Scouting

We found that identification of commercial businesses and technologies appears unsystematic. There is no comprehensive approach in use to identify businesses or technologies; DIOs each appear to have their own processes. More importantly, after business and technologies have been identified, DIOs have no formalized mechanisms to share information relating to interesting technologies with other CTP stakeholders. While the specific technique used for technology scouting can and should vary somewhat based on a number of factors (e.g., DoD customer, nature of the problem to be addressed, phase of the CTP, characteristics of that specific market niche), the resulting information on businesses, their technologies, and the national security problem the technology can address should be documented and shared widely across DoD. Box 5.6 describes an alternative to addressing these issues.

BOX 5.6

Develop More Rigorous, Comprehensive, and Shared Approaches to Technology Scouting

Hypothesis: By developing more rigorous, comprehensive, and coordinated approaches to technology scouting, DoD can leverage the technology scouting activities of other CTP stakeholders.

How the policy lever was evaluated: While we did not directly test this approach in the game, analysis of our interviews suggests that DoD could gain some efficiencies and maximize the value of technology scouting activities by DIOs and other stakeholders by more widely sharing the information generating through these activities across CTP stakeholders.

Links to CTP functions: This approach links to CTP identification functions (technology scouting) and CTP enabling functions (navigation and relationships).

Results

We did not directly test this approach in the game, owing to practical constraints: All players had access to information pertaining to the same set of technologies (though PMs and end users had access to more information on each technology pertaining to requirements). The categories of data we presented to players—description of the technology (including maturity), how it worked, the kinds of DoD problems that could be addressed, DoD and commercial market potential, and vendor description—seemed to provide an acceptable basis for decisionmaking. However, because of the constraints of the game, we did not feel that this collection effectively mimicked the types of information management problems that any real-world database would need to overcome.

Analysis of the interviews that we conducted suggests that DIOs do not share outputs of their technology scouting activities with other DoD stakeholders. Given that these activities are labor-intensive and the outputs may be valuable for other CTP stakeholders, we assess that DoD could gain some efficiencies and maximize the value of these activities by more widely sharing information generated through these activities across CTP stakeholders.

Encourage CTP Stakeholders to Share Problems Across the Enterprise

Many innovation organizations seek to be guided by high-priority end-user problems, but they appear to have their own approaches to sourcing, binning, and cataloging problems. This work is challenging and human capital–intensive. The resulting set of curated problems is a potentially valuable output of DIOs' work, but DIOs do not have a formal or regular mechanism to share problems with other innovation organizations or the DoD. This may result in duplication of effort within and across organizations, mismatches between the problem owner's needs and the available DIO capabilities, and missed opportunities to target novel solutions to problems. If there is no immediate match between a problem and possible solution, the problem, end user, and/or solution may be abandoned. Identification of prospective customers and end users is also essential to businesses, which need access to prospective DoD customers to understand end-user needs and refine potential solutions, but businesses have difficulty locating entry points to DoD and prospective customers. Businesses find it challenging to understand, navigate, and "speak" DoD. Box 5.7 describes an alternative for addressing this issue.

BOX 5.7

Encourage CTP Stakeholders to Share Problems Across the Enterprise

Hypothesis: By encouraging CTP stakeholders at all levels—requirements and capability developers, acquisition organizations (PMs, PEOs, system commands), and especially end users at both the individual unit level as well as major and combatant commands—to share DoD problems across the enterprise, DoD can improve the efficiency of the CTP.

How the policy lever was evaluated: These are core DIO functions, and while we did not test these approaches directly in the game, the game did provide some insight into how these early-stage CTP functions can be improved. Additionally, interviews with DIOs and the background research based on a wide range of official documentation, trade literature articles, and prior research reports suggests that each DIO knows how to perform these functions in their organizational context.

Links to CTP functions: This approach links to CTP core identification functions (problem identification, validation, curation; solution concept generation; customer discovery; matching problems to solutions) and CTP enabling functions (navigation and relationships, coordination, and information sharing).

Results

Fundamentally, success in the identification phase of the CTP depends on the ability of companies and potential DoD end users to identify one another out of a wide universe of information about potential technologies and DoD organizations. By bringing stakeholders together in the game (requiring individuals to opt in to a time-constrained event), that problem was effectively solved for participants. We also provided a limited number of technologies for participants to consider, dramatically simplifying the process of identifying technologies for potential investment. Given this, we did not feel the game alone would be a useful test of policy changes related to these early-stage CTP identification core functions.

However, we did find that widespread information sharing among CTP stakeholders for certain kinds of information—curated DoD problem sets and basic information on technologies and businesses—had a positive effect on collaboration. Simply sharing information about what DIOs were doing (e.g., accelerator programs being conducted, technologies found, problems being worked on) and the capability needs of the acquisition and end-user communities created increased opportunities for collaboration, which therefore increased opportunities to move technologies that addressed a real need through the CTP.

Establish a Single, Comprehensive, Integrated, Searchable Portal for New Entrants

We found that it remains difficult for new entrants to identify opportunities and become known to prospective customers, end users, and acquisition professionals within DoD. For those businesses without a background in DoD, who have limited knowledge of DoD organizational structure and processes, there is no centralized, easily navigable entry point where businesses can identify and easily access DoD opportunities. As a result, DoD is not reaching all innovative technology companies developing innovative technology with potential military applications and may be missing many that could be working with DoD. Box 5.8 describes an alternative for addressing these issues.

BOX 5.8

Establish a Single, Comprehensive, Integrated, Searchable Portal for New Entrants

Hypothesis: By establishing a portal to serve as a common entry point for early-stage ventures that is comprehensive, easy to navigate, and fully searchable, DoD can help new entrants identify and access DoD opportunities, and DoD can leverage the technology scouting activities of other CTP stakeholders.

How the policy lever was evaluated: While we did not directly test these approaches in the game, analysis of our interviews suggests that DoD could help new entrants identify and access DoD opportunities by centralizing and simplifying the process of searching for these opportunities.

Links to CTP functions: This approach links to CTP identification functions (technology scouting, customer discovery, solution concept generation, and network development) and CTP enabling functions (navigation and relationships).

Results

While we did not directly test a centralized portal in the game owing to practical constraints, most of the businesses we interviewed indicated that they found it difficult to identify the problems that DoD needs to address and the office they needed to engage with at DoD to understand and address these problems. Analysis of the interviews we conducted suggests that by centralizing and simplifying the process of searching for DoD opportunities, DoD could help new entrants to identify and access those opportunities.

Establish Navigation Support Services to Help New Entrants Understand and Navigate DoD

We found that businesses have difficulty locating entry points to DoD and prospective customers; they do not understand, know how to navigate, or "speak" DoD. Furthermore, they do not understand the acquisition process or the policy, legal, and contracting aspects of technology development. Some DIOs provide mentoring and navigation support as part of their programming, but many businesses are left asking "What's next?" after DIO programming concludes. These gaps impede technology development and transition. Box 5.9 describes how we evaluated and addressed these issues.

BOX 5.9

Establish Navigation Support Services to Help New Entrants Understand and Navigate DoD

Hypothesis: If DoD established navigation support services, more businesses with promising technologies will identify and apply for opportunities, conduct customer discovery, and understand end-user needs. As a consequence, more technologies will be fielded.

How the policy lever was evaluated: While we did not directly test this approach in the game, analysis of our interviews suggests that by providing additional navigation support, DoD can help businesses understand DoD, identify opportunities, understand "what's next" after completing DIO programming, submit more compelling applications for funding opportunities, and ultimately field more technologies for military use.

Links to CTP functions: This approach links to core functions (customer discovery, solution concept generation, all development and adoption functions) and CTP enabling functions (navigation and relationships).

Results

While we did not directly test navigation support in the game owing to practical constraints, analysis of the interviews we conducted suggests that navigation support could help new entrants more effectively engage with and navigate DoD and ultimately lead to more technologies being fielded. Many of the businesses we interviewed had no prior knowledge of DoD or related funding opportunities, and they did not know how to locate and talk to prospective end users or understand the ways their technology could help the DoD.[3] Most had no idea how to do business with DoD and did not understand the acquisition process or paths for getting on contract.[4] While in some cases these companies' engagement with DIOs helped address some of these challenges, most of the businesses we interviewed noted that they would have appreciated additional navigation support to help them identify a path forward, including funding opportunities to continue development of their technology.

Our interviews with businesses and also representatives of other CTP stakeholders suggest that by providing additional navigation support, DoD can help businesses understand DoD, identify opportunities, understand "what's next" after completing DIO programming, submit more compelling applications for funding opportunities, and ultimately field more technologies for military use.

---

[3] See Appendix B case studies, including Compound Eye and Distributed Spectrum.

[4] See Appendix B case studies, including Compound Eye, Distributed Spectrum, and Vita Inclinata.

Assign Responsibility for Oversight of the CTP, Including Transition Outcomes, to an Organization with the Authority and Budget to Execute Those Responsibilities

We observed that no single organization has responsibility and authority for solving the valley of death and transitioning innovative commercial technology at scale. While most interviewees emphasized that autonomy, agility, and decentralization were required for innovation to occur, we found that the highly decentralized status quo CTP is riddled with gaps and inefficiencies. Box 5.10 describes an approach we evaluated to address these issues.

BOX 5.10

Assign Responsibility for Oversight of the CTP, Including Transition Outcomes, to an Organization with the Authority and Budget to Execute Those Responsibilities

Hypothesis: If a single organization, with the requisite authorities and resources, is assigned responsibility for CTP outcomes, the efficiency and effectiveness of the CTP will be enhanced.

How the policy lever was evaluated: We modeled this in the game by assigning responsibility to an organization. On day 1, this was an existing OUSD/R&E organization taking on this role in addition to its other duties, with only limited authority and no additional funding. On day 2, we created a "DASD(Innovation)" organization within OUSD/R&E with the responsibility and authority to develop enterprisewide innovation guidance and strategy, develop metrics and collect the relevant data, provide oversight and accountability, and manage and allocate the flexible funding pool according to explicit criteria emphasizing collaboration and covering transition gaps.

Links to CTP functions: This approach links to all CTP core and enabling functions.

Results

In day 1 of game play, we found that placing relatively weak centralized oversight responsibility in an existing organization did not appear to affect player decisionmaking. Without real authority and resources, that entity had few effective ways to change other players' behavior to align decisionmaking toward the common goal of moving technology through the CTP and into fielding. However, on day 2, we observed that the relatively stronger authority given to the new DASD(Innovation), together with other policy levers such as stronger policy guidance, game pauses to share information, and the existence of a flexible funding pool focused on transition gaps, resulted in player decisionmaking aligning better toward common goals, improved information sharing, and increased willingness to collaborate with others to transition technology through the CTP. While the white cell played this new organization and its responsibilities with a light touch (trying to balance some degree of centralization against the ability to be agile in execution), it did appear to provide a focus for players and improve CTP outcomes.

We assess that some additional governance and management of the CTP will be required to overcome the identified challenges. Our finding that the disconnect between DIOs and traditional requirements, acquisition, budgeting, and end-user communities adversely affects both CTP throughput and transitions to fielding also comes into play here. This means that the office assigned responsibility for CTP outcomes will need to work closely with counterparts who are responsible for requirements, acquisition, and budgeting processes. Our analysis of this option also highlights the interdependence of many of these options and the importance of implementing these fixes as a coherent package.

Provide Flexible Funding to Incentivize CTP Stakeholder Collaboration and to Close Transition Gaps

The efforts of innovation organizations are to a large extent concentrated in the first set of CTP functions pertaining to identification of technologies, problems, and matching of problems to potential solutions. Fewer innovation organizations have funds to invest in maturing promising technologies. Perhaps the most significant gap we identified in the CTP was the disconnect between innovation organizations, end users, and requirements and acquisition decision authorities. Together, these disconnects hinder handoffs between CTP stakeholders, development of dual-use technologies, incorporation of those technologies in programs, and fielding. Information sharing can partially address some disconnects, as noted above. Box 5.11 describes an alternative for flexible funding that is also needed to resolve disconnects and facilitate technologies moving toward fielding a capability.

BOX 5.11

Provide Flexible Funding to Incentivize Collaboration and Facilitate Transition Gaps

Hypothesis: Flexible funding can serve as a carrot to incentivize CTP stakeholder collaboration and to directly support efforts to close transition gaps.

How the policy lever was evaluated: We tested this policy lever in the game. To set a baseline on day 1, we provided funding to innovation organizations, program managers, and end users and allowed them to use this funding to move technologies into specific CTP activities. We permitted program managers to use some funding with no impact on their current programs and allowed them to request reprogramming of larger amounts of funding, though with some impact on current program schedule. On day 2, we created a flexible funding pool that was equivalent to a 25 percent increase above the day 1 innovation organization funding level. These funds were unrestricted, multiyear funds that were allocated by the white cell, playing the DASD(Innovation). Game participants

---

could request funds for specific purposes to maximize transition opportunities, and particularly the transition from DIO management responsibility to traditional organization responsibility. Per the game construct, requests for supplemental funding were more likely to be approved when a game participant proposed investing their own funds alongside the requested supplemental funding from OSD (i.e., matching funds) and when the proposed activity helped close a transition gap.

Links to CTP functions: This approach links to CTP development and adoption functions and CTP enabling functions: funding and investment, coordination, and plans and planning.

Results

We found that participants were eager to access the additional OSD funds and collaborated and strategized to do so. One participant noted, "I want to tap into other people's money. If the DASD has [extra funding available], I don't want [it] to go to waste."[5] No player requested funds before investing their own funds. Participants requested DASD-allocated funds principally to enable an activity or transition that would otherwise have been difficult to accomplish, but participants had backup strategies if those funds were not approved.

We found that, when program managers were provided with funding to move technologies into specific CTP activities, they were willing to use this "extra" funding for that purpose. However, they were unwilling to take funds from existing programs without end-user support and direction to do so, which no end user provided.[6]

Together with the game's information sharing, coordination, and collaboration mechanisms (described above), which helped address the disconnect between CTP stakeholders, **the flexible funding pool gave players something tangible to collaborate on, helped align players to a collective mission, and helped players move technologies through the pipeline.**

---

[5] RAND NDRI acquisition policy game, March 10, 2022.

[6] It was not clear from game play whether end users did not know this was an option or felt that they could not make that decision at their organizational level.

CHAPTER 6

Key Findings, Recommendations, and Considerations for Implementation

This chapter provides a high-level overview of key findings, offers a suite of mutually supportive and interdependent recommendations to close the gaps and challenges in the CTP that we identified, and suggests considerations for implementation of these recommendations.

Key Findings

The commercial technology pipeline (CTP) is defined as the activities, functions, and processes by which DoD currently identifies, develops, and adopts commercial technologies for military use, including innovation organizations' role in the CTP. Our key findings pertaining to the CTP, and the challenges and gaps that we identified, are explained in the following sections.

Understanding the CTP

- **There is no single "pipeline" or pathway of technologies through the CTP.** The CTP has many potential paths—combinations of activities within and between phases—from concept to fielding, and they are not necessarily linear or sequential. There are many on- and off-ramps at every phase and feedback loops both within and between CTP phases. The path an innovative technology, product, or service takes from idea to fielding can differ depending on characteristics of the technology and business, financial considerations, and alignment with other DoD processes.
- **CTP functions are distributed across multiple stakeholders; no single organization performs them all.** Given this, collaboration and handoffs between CTP stakeholders are essential.
- **Collaboration is required to realize CTP outcomes.** Given the many pathways and distributed nature of the CTP, we assess that collaboration among stakeholders is required to accelerate the identification, development, and adoption of commercial technology for military use.

Innovation Organizations' Role in the CTP

- **DIOs perform multiple CTP functions and use a range of approaches.** Many DIOs concentrate on early CTP functions (the "identification" bin) such as conducting acceleration programs and offering support to develop solution concepts, and they prioritize commercial technologies in defined TRL ranges.
- **DIOs have limited ability to facilitate adoption.** DIOs were established in large part to get new technology into the hands of the warfighter, but they have limited ability to facilitate the adoption of new technology in the final stages: the actual procurement and fielding of technology.
- **DIOs lack consistent buy-in from the traditional communities**, in large part due to incentive structures that are misaligned between DIOs and the traditional acquisition and requirements stakeholders.

Challenges and Gaps in CTP Characteristics

- **CTP stakeholders are not aligned to a shared mission of pushing promising technologies all the way through the pipeline.** DIOs have disparate missions, and while many DIOs view themselves as playing a role in the identification, development, and adoption of commercial technologies, they do not view themselves as part of an integrated system. Many key stakeholders in the traditional acquisition, requirements, budget, and end-user communities do not see themselves as part of the CTP at all. The lack of a shared mission and buy-in from traditional communities ultimately impedes technology transition and contributes to the so-called valley of death problem.
- **DoD has not established goals, objectives, or desired outcomes to guide CTP stakeholders.** There is no DoD-wide strategy or policy guidance for innovation, and other than broad statements of modernization priorities, no specific goals and objectives have been articulated for the CTP. Stakeholders throughout the enterprise each set their own priorities and objectives, which for the most part are not integrated or aligned.
- **CTP stakeholder roles and responsibilities are not well defined or understood.** There appears to be considerable overlap across innovation organization activities and programming, which are mostly focused on early CTP functions, and neither prospective partners in the government nor new-entrant businesses know that many DIOs exist or understand what they do. Support and opportunities for more advanced development activities such as prototypes, live exercises, and demonstrations are distributed unevenly across the department. Traditional acquisition stakeholders, including requirements developers, PMs, PEOs, and end users, do not appear to view themselves as responsible for the integration of dual-use technologies into DoD for military use.
- **Gaps exist in key CTP functions**, as detailed below.
- **No formal mechanisms or requirements exist for information sharing, coordination, or collaboration across the defense innovation ecosystem or CTP.** Rather, information sharing, coordination, collaboration, and feedback mechanisms are ad

hoc and often based on personal relationships. This implies that the CTP and the DIE are not functioning ecosystems, despite observers' frequent use of the term "innovation ecosystem." Ecosystem processes include multiple feedback loops and node interactions that drive how well the ecosystem works. Ecosystems are also self-regulating, usually through some form of information sharing. These conditions are not present today.

- **Incentive structures for CTP stakeholders are not aligned to CTP goals, objectives, and outcomes.** Gaps between innovation organizations, requirements and capability developers, end users, PMs and PEOs, and procurement and decision authorities hinder adoption. Incentives for these relevant CTP stakeholders are not in place to overcome these gaps.

- **No DoD-wide metrics or accountability mechanisms** are in place to check progress against common CTP goals, objectives, and outcomes. Where they do exist within individual organizations, metrics are focused on outputs rather than outcomes.

Challenges and Gaps in CTP Identification, Development, and Adoption Functions

- **Identification of commercial businesses and technologies appears unsystematic.** There is no comprehensive approach in use to identify businesses or technologies, which may mean DoD is overlooking promising technologies.

- **An institutionalized approach to outreach and networking is lacking, and there is no enterprise solution to store related data.** These gaps contribute to duplication of effort, inefficiencies, and missed opportunities. DoD is not capturing the full value of labor-intensive outreach and networking with commercial businesses and DoD partners.

- **DoD end users do not effectively communicate capability needs.** Stakeholders within the same organization identify different priorities, and those priorities shift. Operational units are not aware of other units' problems and are seeking to solve them independently through different channels.

- **Problems are not shared across CTP stakeholders.** Innovation organizations have their own approach to sourcing, binning, and cataloging problems but do not routinely share problems—a potentially valuable DIO output—with other innovation organizations or DoD.

- **Businesses cannot easily locate an initial point of entry into DoD.** There is no centralized, easily navigable entry point where businesses can identify and access DoD opportunities and understand how their technology may apply to DoD missions.

- **Businesses find it challenging to understand, navigate, and "speak" DoD**, including how, where, and when to talk to DoD. As a result, new entrants do not know what DoD needs or how their technology may be relevant.

- **Businesses cannot easily identify the right DoD officials to provide feedback on their technology.** Businesses need to engage prospective end users to understand DoD problems and potential uses of their technology and how to frame and tweak their technology to build a product that will be useful to DoD. However, they have difficulty identifying and becoming known to prospective customers, end users, and acquisition professionals within DoD.

- **There are no warm handoffs, clear routing, or obvious next steps for businesses** once they have come through a particular program or innovation organization. Businesses struggle to understand what is next and to identify subsequent funding and support. Because they have no sustained assistance in identifying next steps, often technology development and adoption stall.

- **Limited suitable funding is available for continued development of technology, and available funding is difficult to locate.** Businesses that are not already familiar with DoD have a difficult time locating information about opportunities on public government websites and portals. DoD has funding available to support development, but much of that funding is concentrated in the early stages of technology development. There is limited support available for continued development of technology beyond a prototype, including testing and proof-of-concept demonstrations, that is needed to help to sustain a company.

- **Most existing opportunities are overly burdensome to pursue.** Burdensome processes are disproportionately costly for new entrants, deterring them from working with DoD. Many small technology companies find the DoD environment, with its insular marketplace and technical and procedural barriers, unappealing and impenetrable.

- **Many factors, including funding, misalignment with the budget process, technical failure, and a lack of formal requirement or customer who is ready, contribute to the "valley of death."** Efforts to address the valley of death problem may fail because they focus on only one of these factors.

- **The valley of death persists in part because no DoD organization has visibility into CTP activities enterprisewide or responsibility for CTP outcomes; and gaps between DIOs, requirements and capability developers, end users, PMs and PEOs, and procurement and decision authorities hinder adoption.** DIOs lack consistent buy-in from the traditional communities, in large part due to incentive structures that are misaligned between DIOs and the traditional acquisition and requirements stakeholders. As a result, fielding ("adoption") functions are gapped almost entirely.

Recommendations to Strengthen the CTP

In making our recommendations, summarized below, we assess that CTP throughput and effectiveness can be enhanced by implementing policy levers that cultivate desired CTP characteristics, encouraging and incentivizing coordination and collaboration, and striking a balance between organizational independence, free-market-style competition, and more centralized direction of the CTP.

Cultivate Desired CTP Characteristics

We recommend that DoD foster the characteristics of a well-functioning CTP that we identified in Chapter 3.

Foster a Shared Sense of Mission Across All CTP Stakeholders

We recommend that DoD identify ways to align CTP stakeholders to a common understanding of the CTP and a shared mission of identifying, developing, and pushing promising technologies all the way through the pipeline. Importantly, key stakeholders in the traditional acquisition, requirements, budget, and end-user communities must see themselves as part of the CTP. By fostering a shared mission, DoD can lay the foundation for information sharing, coordination, and collaboration that is required for the CTP to function. This recommendation can be implemented in a number of ways, ranging from informal guidance from leadership to formal policies and directives.

Develop and Promulgate Strategy, Plans, Policies, and Guidance for the CTP to Establish Common Goals, Objectives, and Outcomes, but Continue to Support Flexibility in Execution

DoD should develop and promulgate DoD-wide strategy, policy, plans, and guidance for the CTP. These issuances should set common goals and objectives and reinforce a sense of shared mission, clarify stakeholder responsibilities, and establish common definitions of key terms and concepts. Working from a common understanding, innovation organizations and other CTP stakeholders will be encouraged to pursue problems and technologies that are deemed to be of highest priority by leadership and to share information and collaborate more readily. This is not to say that the operation of the CTP should be centrally directed; DoD should continue to support flexibility and agility in execution of department-level strategy, policy, plans, and guidance.

As a foundation, we recommend that "success" for the CTP—the goals of the overall system—be clearly defined, communicated, and disseminated across CTP stakeholders in such a way that they are understood and shared by all stakeholders. These goals could include the following:

- Fielding technology to the warfighter, which covers maturing promising technology that leads to some form of adoption (fielding residual capability, limited fielding, and fielding at scale).
- Maturing promising technology that is taken up by commercial industry and increases U.S. competitiveness, on the assumption that economic success underpins national security.
- Expanding the industrial base to ensure it is responsive to DoD capability needs.
- Maximizing opportunities for adoption through collaboration.

To support stakeholder collaboration, guidance should provide clarification on key issues. These issues include stakeholder roles and responsibilities for CTP outcomes and key terms, like "transition," that set expectations and establish mechanisms for information sharing, coordination, and collaboration, establish outcome-related metrics, and establish account-ability and oversight mechanisms—points that we return to below.

Define and Communicate Roles and Responsibilities for the CTP, Ensuring There Are No Gaps in Key Functions

DoD should clarify and formalize roles and responsibilities for CTP stakeholders, specifying which entities are expected to perform which CTP functions. DoD should also clarify and formalize the relationships between these stakeholders. Critically, DoD must ensure that traditional acquisition stakeholders, including requirements developers and program managers, understand their roles and responsibilities for CTP outcomes as defined by the department and that they view these roles as consistent with and a distinct part of their job. Given the large number of innovation organizations and the overlap in what many are seeking to do, DoD may want to consider optimizing the number and focus of DIOs to minimize duplication and overlap and fill gaps. It is also important to acknowledge that some overlap in functions, technologies, and customer focus among DIOs is useful to sustain because not every innovative commercial technology can be expected to succeed.

Facilitate Information Sharing, Coordination, and Collaboration Across CTP Stakeholders

DoD should establish mechanisms to improve information sharing, coordination, and collaboration across innovation organizations and other CTP stakeholders. For example, we recommend that CTP stakeholders share and institutionalize information relating to end-user problems and needs, new-entrant ventures, technologies and their potential applications, investment considerations and decisions, transition plans, and contacts both inside and outside DoD.

There are many ways to implement this approach, some more centralized than others. Simply making these communities aware of each other's activities and needs, perhaps through a simple shared database that captures relevant information, would increase the opportunities for collaboration and reduce potential nonvalue-added duplication of effort. DoD might also consider improved information systems, reporting requirements, headquarters-led information sharing across CTP stakeholders, and recurring meetings. In addition, DoD could encourage information sharing, coordination, and collaboration in its policy guidance. Any of these additional requirements should be analyzed and implemented carefully to ensure that the cost-benefit ratio justifies any additional burden on stakeholders.

Establishing mechanisms and processes to support these activities is only one component of the changes that will be required. To establish the conditions under which *value-added* information sharing, coordination, and collaboration will occur and be effective, DoD must also foster a sense of shared mission and shift incentives for CTP stakeholders. We return to this point and also the challenge of broader cultural change below as we discuss considerations for implementation.

Implement Incentive Structures, Including Metrics and Accountability Mechanisms, to Align CTP Stakeholders to CTP Goals, Objectives, and Desired Outcomes

DoD should realign incentive structures for all CTP stakeholders, including but not limited to innovation organizations that are aligned to CTP outcomes. This effort should include developing meaningful metrics tied to the priority CTP outcomes that are specified in strategy and guidance, as well as accountability and oversight structures to monitor compliance with the written guidance. This package could also include a flexible funding pool available to CTP stakeholders that would operate as a "carrot" to incentivize stakeholders to work toward CTP outcomes (we discuss this suggestion below). As part of this package, DoD could train and incentivize acquisition professionals to make better use of existing authorities and mechanisms for engaging with early-stage ventures and their commercial technologies and to utilize OTA and other authorities that facilitate adoption of commercial technologies.

In developing metrics, accountability structures, and other incentives, we recommend a focus on more than just fielding. For example, DoD could encourage CTP stakeholders to take the following actions:

- Build partnerships with other CTP stakeholders, regularly seeking their input on prospective investment decisions.
- Develop and utilize warm handoffs and pathways to advance technologies through the pipeline.
- Leverage previous SBIR investments by incentivizing stakeholders to look to databases of previous SBIR awardees to fill capability gaps.
- Foster and/or directly hire identified talent into DoD to develop innovative capabilities, where doing so does not compete with other CTP goals.

These examples are intended to encourage more engagement among CTP stakeholders, particularly between the DIOs, and between the DIOs and the traditional requirements, acquisition, budgeting, and end-user communities.

One specific metric that DoD should consider is throughput—the number of new technology ideas or activities or opportunities created in or moving through the CTP per unit time. This metric places value on all transitions or handoffs occurring with the CTP, not just the final fielding metric. Higher throughput also increases the opportunities for adoption. Because it covers the entire innovation life cycle, it allows all CTP stakeholders to contribute and measure their own performance with respect to their contribution to a shared objective (maximizing throughput of commercial technology opportunities) while also providing some insight into CTP performance and outcomes. A viable commercial business is an important intermediate outcome of the CTP and contributes to ensuring DoD's ongoing access to commercial technology. The number or rate at which viable new commercial businesses enter the defense industrial base is therefore another potential enterprise-level metric.

These metrics should be designed to encourage collaboration and mitigate the factors that result in the valley of death phenomenon. By shifting incentives, DoD can maximize the likelihood that innovative commercial technologies move all the way through the pipeline.

Strengthen CTP Identification, Development, and Adoption Functions

The following recommendations are designed to strengthen and fill gaps in specific CTP functions.

Develop More Rigorous, Comprehensive, and Coordinated Approaches to Technology Scouting

We recommend that DoD develop more rigorous, comprehensive, and coordinated approaches to technology scouting. This could include more coordinated outreach efforts such as road shows or participation in regional-based events. Activities include advertising in local, regional, or national media; publishing requests for information and requests for proposals; pushing email notifications to universities, research organizations, and established commercial businesses; and establishing a physical presence at regional innovation hubs. The information on technologies and businesses developed as a result of these activities should be widely shared across the DoD enterprise.

Encourage CTP Stakeholders to Share DoD Problems Across the Enterprise

We recommend that DoD encourage CTP stakeholders at all levels—requirements and capability developers, acquisition organizations (PMs, PEOs, system commands), budget offices, and especially end users at both the individual unit level as well as major and combatant commands—to share DoD problems across the enterprise. For example, DoD could establish a repository of DoD problems and common terminology describing those problems and encourage its use by innovation organizations, services, and combatant commands (CCMDs) as end users. This will allow DoD to realize the value of problem curation—a labor-intensive and potentially valuable output from DIOs.

Establish a Comprehensive, Integrated, Searchable Portal for New Entrants to Access DoD Opportunities

DoD should establish a portal to serve as a common entry point for early-stage ventures (e.g., "If you are early-stage, click here"), building on the OSD innovation website.[1] This website lists a subset of innovation organizations, funding opportunities, and guidance for small businesses, but it appears to be only a subset of the information on organizations, processes, and opportunities that may be relevant to new entrants. We recommend augmenting the website so that it is more comprehensive, easier to navigate, and fully searchable (e.g., where the contents of the linked websites of organizations and opportunities can be searched).

---

[1] OUSD/R&E, undated a.

Through this portal, DoD could explain how commercial technologies can be identified, developed, and adopted (providing examples of different pathways); expound on different types of opportunities; and describe what each of the CTP stakeholders does as part of the pipeline. The portal could also be a vehicle to collate and publicize funding opportunities to support concept and technology development and opportunities to engage prospective users and program managers, and it could provide guidance on the types of information that companies should be prepared to share with the department.

Establish DoD Navigation Support Services to Help New Entrants Understand and Navigate DoD

DoD should establish navigation support services to help ventures with promising technology understand DoD, navigate and apply for opportunities, understand end-user needs, and conduct customer discovery. Many of the DIOs do provide some initial guidance through accelerator and other similar programs, but even after that initial touchpoint, many businesses need continued help navigating the DoD bureaucracy and the policy, legal, and contracting aspects of technology development. A navigation capability would also address the "What's next" issue that was identified as a challenge by many interviewees. There are currently few "warm handoffs," clear routing, or obvious next steps for businesses once they complete an accelerator or other similar DIO program or initiative. As noted above, DIOs do not perform this function in the later phases of the innovation life cycle. These handoff points, especially in the development phase, become small valley of death barriers. By providing navigation support, DoD can enable more businesses with promising technologies to identify and apply for opportunities, conduct customer discovery, and understand end-user needs; and as a consequence, DoD may field more technologies.

Assign Responsibility for Oversight of the CTP, Including Transition Outcomes, to an Organization with the Authority and Budget to Execute

We recommend that DoD assign responsibility for CTP oversight and realizing CTP outcomes, including transition, to an organization. This organization should be provided with the authority and budget to execute these responsibilities, including developing and promulgating strategy, guidance, and plans; developing and implementing metrics and accountability mechanisms; and facilitating information sharing, coordination, and collaboration across CTP stakeholders. This organization would also manage the pool of flexible funding described below.

Viewed holistically, the CTP (and broader DIE) is really just part of the acquisition process. In fact, the CTP can be thought of as similar to the distinct pathways defined within the Adaptive Acquisition Framework (AAF). That means that the office assigned responsibility for CTP outcomes will need to work closely with counterparts who are responsible for requirements, acquisition, and budgeting processes.

Assigning responsibility for governing the CTP will give an organization ownership of the process and outcomes and facilitate monitoring performance and making changes in policy and process to improve outcomes.

Provide Flexible Funding to Incentivize CTP Stakeholder Collaboration and Close Transition Gaps

We recommend that DoD establish a flexible funding pool to facilitate transitions and support the maturation of promising technologies (that might otherwise stall in the valley of death) and the incorporation of those technologies in a program of record where appropriate. While funding availability is only one factor in the valley of death, it is an important one. By flexible funding, we mean funds that can be applied to any development or early production activity (i.e., not restricted by "color of money" issues), are not tied to the normal PPBE cycle (i.e., not programmed for a specific program or system but rather for a general pool to support commercial technology innovation and integration), are not tied to an approved requirement, and are controlled and allocated by the single organization assigned responsibility for the CTP.

In allocating this funding, DoD should prioritize support for CTP activities that have a significant impact on fielding, such as prototype demonstrations, and participation in live exercises designed to enable evaluation of military utility. This funding pool could also help to sustain a business and its technology, providing funds to fill the gap where an engaged customer is working to include a technology in the next budget cycle, or it could be used to field a technology or product as a residual capability. To access these funds, CTP stakeholders could be required to provide matching funds, which would ensure that these stakeholders had a high degree of interest, engagement, and commitment.

DoD will need congressional approval to implement this recommendation. A flexible funding pool that can supplement the existing budgets of CTP stakeholders will only be effective if Congress provides adequate funding. Additional analysis is required to estimate the level of funding needed, which may vary over time, and to develop allocation criteria and distribution processes.

The combination of more widely sharing information on DoD problems and candidate technologies, enhanced navigation assistance, assigning ownership of the CTP processes and outcomes to a single office with requisite authority to establish goals and incentivize collaboration, and the flexible funding pool is intended to directly address the valley of death phenomenon.

Considerations for Implementation

Recommendations Are Mutually Supportive, Interdependent, and Should Be Appropriately Calibrated

Our recommendations are not independent of each other. They are designed to be interdependent and mutually supportive and should be implemented as a package. For instance, we have recommended that DoD issue policy guidance to define the desired outcome(s) of the CTP. This guidance should also identify CTP stakeholders and clarify their roles and responsibilities and establish (or assign) an organization to be responsible and accountable for ensuring the CTP runs as smoothly as possible and produces the desired outcomes. The organization assigned that responsibility should develop and promulgate an enterprise innovation strategy and develop metrics that provide insight into CTP performance and outcomes. That strategy should also include specific measures addressing and mitigating the factors that result in the valley of death phenomenon.

Many of our recommendations are also intended to address more than one of the challenges we identified. For example, by establishing a mentoring or navigation function, DoD can help to address several of the challenges that new entrants face: too many entry points to the department, difficulty accessing prospective customers for customer discovery, and the lack of identified next steps and warm handoffs between DIOs, CTP functions, and phases.

CTP enabling functions—which drive activities in each of the three innovation life-cycle phases—are not occurring regularly or are occurring on an ad hoc or localized basis across CTP phases. We recommend that DoD strengthen these enabling functions by implementing the recommendations described above. Importantly, these governance tools must be calibrated to preserve the agility of innovative businesses and DIOs and increase the agility of the traditional processes responsible for bringing new capabilities to the warfighter.

Implement DIO Best Practices

While DIOs cannot solve these problems on their own, DIOs should implement the following best practices in their own work:

- Be led by a DoD problem rather than by innovative dual-use technologies. There appears to be a consensus that success (i.e., a technology being fielded at some level) is more likely if the technology search works from actual user capability gaps or specific problems of operational units. The alternative—identifying an innovative commercial technology and then searching for a problem it can solve—is considered a less successful approach.
- Include transition planning as early as possible, even during accelerator or incubator programs, and revisit that transition plan continually. This includes identifying and engaging potential transition partners early.
- Recognize that an engaged DoD customer is critical for success. An engaged customer can participate in problem curation, selecting a cohort for an accelerator, and may have direct interaction with ESVs helping to refine potential solution concepts. This early engagement is particularly helpful to ESVs, though it is also helpful, of course, if the potential customer/transition partner has budget authority and access to contracting.[2]

Entrepreneurial Persistence and Creativity Is Required by All CTP Stakeholders

One condition for implementation success—executing the CTP well—concerns individual stakeholder characteristics. A common and recurring theme of acquisition case studies, including those focused on innovative technology and capabilities, is the importance of

---

[2] Interview 8, Interviewee 9, August 16, 2021.

knowledgeable, experienced, risk-taking personnel in key positions who know how to work the system. In the context of technology innovation and adoption, entrepreneurial persistence and creativity and ginning up demand for the product through demonstration are important factors affecting success. Having the right people in the right place with the right knowledge and the motivation and drive to make something happen still factors heavily in successful throughput and transitions. This condition is needed across CTP stakeholders.

Further Research

The CTP and, relatedly, the requirements and acquisition processes are nuanced, and changes to any part of these systems will require additional study. Here we identify some of the areas that will require additional focus as DoD considers reforms.

Assess the Costs, Benefits, and Unintended Consequences of Tweaks to This Complex System

The operation of the CTP is complex, with many stakeholders in different organizations, each contributing functions and activities necessary to identify, develop, and adopt commercial technology for military use. Any effort to tweak this complex system may introduce additional complexities and nuances, and DoD should carefully assess the costs, benefits, and unintended consequences of each potential change—an examination that was beyond the scope of this study. Toward this end, the CTP model provides a framework that DoD can use to establish a common understanding of the CTP and how it works, to standardize definitions of key terms, to help innovation organizations and other CTP stakeholders in assessing their role(s) and understanding the roles of other stakeholders in that pipeline, and to assess tweaks to the system.

Continue to Explore and Understand CTP Pathways and Their Value to DoD

While DoD has long relied on the private sector for innovative technology, two CTP pathways should be of particular interest to DoD. In one, DoD uses the commercial marketplace to mature technologies that can be used by DoD to solve military problems or address a particular need. This can include technology development initiated in the DoD (or government) lab structure and licensed or otherwise provided to one or more commercial businesses. In the second, the military becomes an early adopter or funder to get new ideas into the commercial market, where it can then mature while continuing to support the military need. These are both specific CTP pathways that make use of the commercial market to mature and refine technology and products, as well as sustaining the viability of the business.

Identify Additional Ways to Close the Gap Between DIOs and Traditional Requirements, Acquisition, Budgeting, and End-User Communities

Many of the core and enabling functions necessary for a technology to progress through the CTP are the responsibility of the requirements, capability development, acquisition, budgeting, and end-user communities. We observed that DIOs are disconnected from these communities and assess that this disconnect remains the principal constraint on both CTP throughput and transitions to fielding. Integrating these traditional requirements, acquisition, budgeting, and end-user stakeholders more effectively in the CTP, especially in the identification phase, would increase commercial technology opportunities in the development and adoption phases. Although we have made recommendations to close this gap, we suggest that DoD also examine other ways to do so.

Leverage Acquisition Policy Gaming to Test Policy Levers

The acquisition policy game developed for this study is an output of the research. It provides a platform to test alternative approaches to strengthen and accelerate the adoption of commercial technology for military use. We tested several policy levers in this research and gained insight into others, but there are many other policy levers that we did not address here. The existing game platform can be modified and improved to include additional stakeholders or test other policy levers.

Explore Ways to Support Cultural Adaptation and Change to Maximize CTP Outcomes

There is a growing awareness that adopting innovative commercial technology for military use is not just a technical problem but rather a cultural one. Technical problems can be defined and fixed by identifying a technical fix—a technology that accomplishes the needed task. But innovative technology often requires adaptation—behavioral changes in how a problem is addressed.[3] The willingness to make the process and organizational changes that may be required to make use of an innovative commercial technology is often a cultural issue and can constrain or hinder adoption of that technology. DoD should explore this aspect of technology adoption in order to improve CTP outcomes.

---

[3] Brandon Leshchinskiy and Andrew Browne, "Digital Transformation Is a Cultural Problem, Not a Technological One," *War on the Rocks*, May 17, 2022.

Defense Innovation Organization Deep Dives

Army Applications Laboratory

Snapshot of organization	**Founded:** Following the stand up of Army Futures Command (AFC) in summer 2018, AFC innovation officer Adam Jay Harrison was appointed to establish AAL as a subsidiary organization.[a]

Mission and focus: AAL is intended to act "as a concierge service across the Army's Future Force Modernization Enterprise and the broader commercial marketplace of ideas."[b] AAL's mission is to "align innovative solutions and technologies with Army problems, resources and programs to rapidly discover, validate and transition technology applications in support of Army modernization."[c] Key areas of focus and priority technologies include maintenance and logistics (robotics, autonomy); soldier performance through training and augmented reality/virtual reality (AR/VR); communications and information (cybersecurity, C3I components, networks); energy and power (low-power electronics, compact power sources); GPS-free navigation (PNT and algorithms, space); and situational awareness and real-time intelligence including sensors, sensor fusion, and unmanned artificial intelligence/ machine learning (AI/ML).

Organizational statistics: Annual budget of approximately $66 million.[d]

How it operates: AAL performs three core functions: market intelligence (market research, trend analysis, commercial assessments); solution evaluation (technical assessments, proof of technology, solution validation); and solution innovation (problem framing, concept development, solution demonstration, transition support).[e] AAL ultimately seeks to match Army sponsor organizations with companies/technologies through a process of problem identification (either solicited by an AFC cross-functional team or directly curated from an operational unit) and market intelligence.

AAL's SPARTN program—Special Program Awards for Required Technology Needs—uses three distinct approaches to match problems with solutions through the SBIR process: *area* ("unknown unknowns"); *cohort* ("system of systems"); and *point* ("a specific solution, tailored to meet a detailed problem statement").[f]

- The *cohort* approach supports up to 15 businesses for one to three years. It is designed to be a "flexible, collaborative environment to solve a multifaceted [specific Army] problem."[g] Companies receive $200,000 in phase I and up to $2.5M if selected for phase II. Awards are sequenced following the traditional SBIR process (phase I followed by phase II and potentially phase IIB).
- The *point challenge* approach supports up to five businesses for one to two years. Companies are "separately tasked to develop technology tailored to a specific problem statement."[h] Funding and periods of performance depend on the specific context of a given problem statement. Funding is flexible and can be tailored to the specific company response and technology development stage (i.e., a company could either receive phase I or direct-to-phase II awards).

- The *area challenge* supports up to eight businesses for one to two years. Companies are "separately tasked to develop technology to solve a wide-ranging problem."[i] Funding and periods of performance depend on the specific context of a given problem statement. Funding is flexible and can be tailored to the specific company response and technology maturity level (i.e., a company could either receive phase I or direct-to-phase II awards).

According to an interviewee, AAL assists Army customers in refining the problem statement, then it works with them to "identify funding source[s] . . . SBIRs, OTAs, etc.; identify stakeholders; [find] the PM, contracting, funds; [line] up dollars, authorities, [and] capabilities to take to fielded environment before solicitation."[j] AAL then chooses between two and 15 companies based on initial submissions, in accordance with statutory timelines.[k] Before solicitation, AAL brings in stakeholders for meetings and technical calls to gather technical papers that are ultimately reviewed for selection.[l]

Selection notification takes about 15 days. Finally, the time from close of solicitation to contract is about 45 days (compared with 280 days in typical Army contracting). In the cohort program, for example, a cross-functional team will identify a modernization priority, which AAL will advertise using social media and other digital communications. After selecting a small group of applicant companies, AAL sends the participants to learn about the problem directly from end users. Over the course of about 12 weeks, participants work to develop a concept or solution that may eventually inform requirements or result in a follow-on contract with an Army sponsor.[m]

Primary customers: U.S. Army.

Link to CTP	**Identification functions performed:** AAL performs technology scouting; problem identification, validation, and curation; solution concept generation; network development; and matching problems to potential solutions.
	Development functions performed: AAL performs further development and maturation of solution concept, technology maturation (early stage), and RDT&E activities (including prototyping, user testing, and feedback).
	Adoption functions performed: AAL performs transition to fielding by assisting companies get on contract with an Army sponsor.
	Enabling functions performed: AAL performs coordination, funding and investment, navigation and relationships, and data collection and analysis.
How they do their CTP functions	**Technology scouting:** AAL conducts market intelligence and technology scouting.[n] AAL also announces all new opportunities through the organization's social media and website platforms, as well as through a curated mailing list of interested companies.[o] According to interviewees, "Once we scout the problem, we look at the marketplace," beginning with corporate ventures.[p] "We look at the state of the industry that exists around that problem; this is liable to include data fabric, AI, robotic arms, or some new tech. What is the tech stack? Where does that stack live? Look around the country. There are areas around the country based on universities or companies. What is the maturity of [the] industry—early or late stage?"[q]
	Problem identification, validation, and curation: AAL identifies problems through two principal channels: cross-functional teams (CFTs) and operational units. First, CFTs that are intrinsically aligned with Army modernization priorities and are "focused on specific areas—fires, soldier lethality, PNT (position, navigation, and timing), various other things . . . [they] bring people together and say we have this problem."[r] According to interviewees, AAL PMs "are in contact with CFTs on a regular basis." The advantage of this approach is that it fosters a close linkage with what the broader Army is doing; however, a disadvantage is that it necessarily precludes any activities that fall outside the CFTs' purview. Second, AAL "send[s] people to talk to [operational] groups—82nd, etc." and asks, "What are you trying to solve?"[s] If the problem set is not refined—that is, it is "incredibly broadly defined, [for example] we want to do something with electrification"—AAL will help to refine the problem by connecting with contacts at universities and other researchers.

Solution concept generation: AAL helps small businesses develop and refine solution concepts through two primary functions: solution evaluation (technical assessments, proof of technology, solution validation) and solution innovation (problem framing, concept development, solution demonstration, transition support). Through the SPARTN program, AAL selects a group of companies to participate in one of three SBIR approaches (described above): cohort, point challenge, and area challenge. As noted above, AAL sends the SPARTN participants to learn about the problem directly from end users and works with the participants to develop a concept or solution.

Network development: AAL has established extensive relationships throughout the DIE with other defense innovation organizations, Army customer bases, subject-matter experts (SMEs), and innovators in academia. AAL's outreach and communication with other DIOs has primarily centered on identifying potential funding sources. An interviewee noted: "There are other options; we've been working with other innovation units—DIU, AFWERX, NavalX—to find multiple levels of funding to help while we're pursuing POM [program objective memorandum] funds. We're finding other funding sources."[t] As part of its purview, AAL maintains several communication channels with Army CFTs and SMEs (see below). Finally, AAL curates relationships with university researchers to help Army customers refine their problem statements. In addition to maintaining contact with other DIOs to share resources and funding opportunities, AAL has accumulated a database of private companies and entrepreneurs that it uses as a mailing list to send out new Army opportunities and competitions.

Matching problems to potential solutions: AAL's matching process includes both the SPARTN program and RDT&E-funded projects. It connects nontraditional companies/technologies with cross-functional teams, sponsor organizations, and organizational units. Following the announcement of a new opportunity in the SPARTN program, AAL reviews submissions and brings in stakeholders before solicitation to discuss the relevance of each submission to potential problem statements. Finally, AAL works with the Army customer to down-select companies for a SBIR phase I or II award or another contract vehicle.

Further development and maturation of solution concept: See above, *Solution concept generation*. As part of the SPARTN program, AAL connects companies and entrepreneurs with Army end users to refine solution concepts, ultimately leading to a follow-on contract or prototype/demonstration effort.

Technology maturation (early stage): The SPARTN program includes three principal approaches to early-stage technology maturation. The *cohort approach* supports up to 15 businesses for one to three years, sequenced through SBIR phase I/II. The *point challenge approach* supports up to five businesses for one to two years, with SBIR phase I and II flexibly applied to the given problem statement. The *area challenge approach* supports up to eight businesses for one to two years, with SBIR phase I or II again flexibly applied. Additionally, AAL facilitates early-stage prototyping and demonstration efforts through traditional Army RDT&E funds.

RDT&E activities: AAL uses programmed RDT&E funding to support prototyping, technology development, and demonstration up to prototype (TRL 6).[u]

Transition to fielding: AAL assists companies and sponsor organizations in transitioning technology to a follow-on contract, program of record, or contractual vehicle. AAL uses three primary transition types to support nontraditional companies' proof-of-concept, prototype, or demonstration efforts: (1) directed subcontract under a prime contractor, (2) a patchwork of funding leading into a POM submission, and (3) dual use.[v] According to interviewees, once a company has generated a prototype and is waiting to receive RDT&E funding through the POM process, AAL can help sustain the company through the Rapid Defense Experimentation Reserve fund.[w] Additionally, AAL supports companies with a strong use case to pursue dual use for commercialization, particularly if they already possess a robust commercial revenue base. For companies or entrepreneurs who are not able to sustain a full dual-use production, AAL also fosters introductions to outside businesses for possible directed subcontract opportunities. It appears that AAL may occasionally offer funding to companies that seek to pursue dual-use opportunities, but this needs to be clarified.

Coordination: See above, *Network development.*

Funding and investment: AAL finds funding opportunities (RDT&E funds, SBIR/STTR, OTAs, etc.) to enable small businesses to develop prototypes, proofs-of-concept, and technology demonstrations. Additionally, AAL leverages its network of DIOs to share funding opportunities and resources.[x]

Navigation and relationships: See above, *Network development.* The SPARTN program provides participating companies and entrepreneurs with direct access to Army end users and SMEs to offer feedback on the participants' solution concept development.

Data collection and analysis: AAL provides a variety of data analyses to both Army customers and commercial entities, including market research, trend analysis, commercial assessments, technical assessments, and solution validation.

CTP-related outputs	AAL's matching process includes both the SPARTN program and specially allocated RDT&E projects. SPARTN's principal outputs are SBIR phase I/II agreements, while AAL's RDT&E-funded projects yield prototyping and demonstration efforts under RDT&E program elements, directed subcontracts to primes, and other contracting solutions (e.g., OTAs).[y] Additionally, AAL provides a variety of intermediate outputs to both Army customers and commercial entities including market research, trend analysis, commercial assessments, technical assessments, and solution validation.
Definition of success	According to interviewees, "a win is a transition," though "not necessarily into a program of record."[z] Transition could also mean adoption of the technology by another unit or government agency, potentially for a different purpose: "A unit [may say] we'll only buy 60, but we have a partner that takes it off our hands and goes running with it. Success is also that sixth company; we'll put them on a phase III unrelated to our company."[aa] AAL views as its ultimate charge "creating best practices that help small businesses" and help the Army work with more small businesses.[bb] Success, therefore, is fundamentally about successfully enabling a small business to field its solution with an Army unit.
CTP-related outcomes	According to AAL, nearly two-thirds of AAL projects transitioned to operational units in 2020, in part due to the start of AAL's new SPARTN program.[cc] Success stories include a start-up that developed and demonstrated a robotic arm for autonomous ammunition resupply and a small business that developed a real-time cannon ammunition management system.

NOTES: This profile and others in Appendix A are constructed primarily from original interview comments and public information available on various websites.

[a] Army Futures Command, "AAL," webpage, undated.

[b] Army Futures Command, undated.

[c] Army Futures Command, undated.

[d] AAL Joint DODx Update Group Brief, obtained by authors through communication with AAL, undated.

[e] AAL, "What We Do," webpage, undated e.

[f] AAL, undated e.

[g] AAL, "SPARTN Info Sheet," webpage, undated d.

[h] AAL, undated d.

[i] AAL, undated d.

[j] Interview 17, Interviewee 19, September 24, 2021.

[k] Interview 17, Interviewee 19, September 24, 2021.

[l] Interview 17, Interviewee 19, September 24, 2021.

[m] AAL, "Case Study: The Field Artillery Autonomous Resupply (FAAR) Cohort," Info Sheet, webpage, undated b.

[n] AAL, undated e.

[o] AAL, "Get Involved," webpage, undated c.

[p] Interview 17, Interviewee 19, September 24, 2021.

[q] Interview 17, Interviewee 19, September 24, 2021.

[r] Interview 17, Interviewee 19, September 24, 2021.

[s] Interview 17, Interviewee 19, September 24, 2021.

[t] Interview 17, Interviewee 19, September 24, 2021.

[u] AAL, undated e.

[v] Interview 17, Interviewee 19, September 24, 2021.

[w] Interview 17, Interviewee 19, September 24, 2021.

[x] Interview 17, Interviewee 19, September 24, 2021.

[y] Interview 17, Interviewee 19, September 24, 2021; also see AAL's website.

[z] Interview 17, Interviewee 19, September 24, 2021.

[aa] Interview 17, Interviewee 19, September 24, 2021.

[bb] Interview 17, Interviewee 19, September 24, 2021.

[cc] AAL, "About the SPARTN Program," webpage, undated a.

AFWERX

Snapshot of organization

Founded: AFWERX officially launched July 21, 2017, by then-SecAF Heather Wilson, based on a model used by SOCOM, to better connect universities, small businesses, and entrepreneurs to the Air Force.[a] In 2019, the SBIR/STTR Center of Excellence was added to the AFWERX enterprise, and employees began referring to the organization as AFWERX 2.0.[b]

Mission and focus: AFWERX is a program office within the Air Force Research Laboratory (AFRL) that aims to develop and expand the adoption of new technologies at all levels of the U.S. Air Force (USAF) and U.S. Space Force (USSF) by connecting innovators across academia, government, and industry.[c]

Organizational statistics:

- In FY 2022, AFWERX's revenue was appropriated at nearly $28.8 million.[d]
- AFWERX's first location was opened at the University of Nevada, Las Vegas; and it has since opened innovation hubs in Austin, Texas; Washington, D.C.; and a SpaceWERX hub at Los Angeles Air Force Base.[e]
- Spark cells, groups of airmen who collaborate to generate ideas and projects locally, exist worldwide.[f]

How it operates: AFWERX has three main lines of effort: AFVentures, Spark, and Prime. A fourth, SpaceWERX, was launched in 2021 as the Space Force–affiliated arm of AFWERX. All these efforts are meant to spark innovation across the Department of the Air Force (DAF) by tapping into commercial tech ecosystems and accelerate the adoption of commercial market tech products.[g]

- *AFVentures* acts as a commercial investment arm facilitating DAF investments in commercial industry to find and develop disruptive technologies. This venture manages DAF's SBIR/STTR program called Open Topic, which leverages a portion of the SBIR/STTR budget to enable the Air Force to develop and adopt commercially viable innovations while providing a competitive edge to the U.S. entrepreneur and technology system.[h] The Open Topic model includes solicitations, phase II funding, contracting and execution, and strives to find supporting acquisition organizations or MAJCOMs within the DAF to adopt potential solutions and transition them to the warfighter.[i]
- *Spark* seeks to change the culture of the services to better support innovation by allowing collaboration between airmen, guardians, and commercial innovators to solve problems at the local level. Spark cells are groups of airmen at different units and bases worldwide who can generate and execute ideas and projects at their level. An annual Spark Tank event allows airmen and guardians to pitch ideas to DAF leadership and industry experts.[j]

- *Prime* is an accelerator program that can leverage DAF resources to accelerate the development of nascent commercial technologies. Prime focuses on taking technologies that have matured through SBIR/STTR (or other investment opportunities) and finding the rapid acquisition or transition pathway to further tailor or develop a solution for the warfighter.[k] One interviewee noted that Prime exists to get into technologies that have matured beyond the basic S&T stage and have already created a demonstrated capability to bring them into DAF. Under Prime, initiatives focus on specific technologies: Agility Prime focuses on electric vertical takeoff and landing (eVTOL)technologies; Vector Prime focuses on supersonic capabilities; and Autonomy Prime focuses on autonomous technologies.[l]
- *SpaceWERX*, a new USSF-affiliated arm of AFWERX, focuses on the development and adoption of technologies that could be used by USSF. It was announced in December 2020 and officially launched in August 2021 based at Los Angeles Air Force Base.[m] AFWERX jump-started SpaceWERX initiatives through SBIR phase I Open Topic solicitations on space technologies.[n] SpaceWERX has its own .us website with programs that mirror AFWERX (Venture, Spark, and Prime), but it is unclear from website descriptions whether these programs operate the same way as AFWERX programs.[o]

Primary customers: Department of the Air Force: USAF and USSF.

Link to CTP	**Core functions performed:** Technology scouting, customer discovery, solution concept generation, network development, matching problems to potential solutions, further development and maturation of solution concept, S&T activities (includes experimentation), technology maturation (early stage). **Enabling functions performed:** Funding and investment, coordination, hard infrastructure to support development.
How they do their CTP functions	**Technology scouting:** Through AFVentures, AFWERX seeks solutions to USAF and USSF problems across the commercial industry and acts as an investment arm for DAF. AFWERX's goal is to find disruptive technologies to both shore up the industrial base and keep the USAF and USSF competitive. AFWERX scouts in a few ways, including soliciting applications for SBIR/STTR investments and issuing challenges to academia and industry to create solutions to problems. AFWERX, as part of AFRL, is also able to get a sense of what technologies are under development at the lab and can find solutions there that may be of use to the DAF but that AFRL on its own may not be able to move forward.[p] Through Prime, AFWERX is able to search for technologies on specific topics and solicit applications from ventures that have matured past the basic R&D stage to a demonstrated capability. These more advanced technologies are then further developed through Prime to become commercially viable and useful to the military; one interviewee noted the ideal timeline for a Prime-sponsored project would be 36 to 48 months to accelerate a technology solution to a point where it is useful for the warfighter.[q] **Customer discovery:** Primarily through AFVentures and Prime, AFWERX does "matchmaking" by facilitating introductions between industry, academia, operators, end users, and traditional acquisition customers.[r] This matchmaking is intended to facilitate conversations and relationships, as well as increase transition opportunities. **Solution concept generation:** Through Sparks, AFWERX empowers operators at the base level to collaborate among one another and with industry and academia to solve problems, allowing for rapid communication and scaling. Using SBIR/STTR, AFWERX solicits solutions to a variety of problems and invests in early-stage development of new technologies and/or processes. AFWERX reviews companies in phase II to seek other DAF organizations willing and able to sponsor a developing technology or capability to increase its chances of transition.[s] **Network development:** AFWERX encourages communication and interaction between academia, industry, operators, and traditional acquisition offices to facilitate ongoing collaboration.

Matching problems to potential solutions: AFWERX does this matching through management of the SBIR/STTR process. Through Sparks, AFWERX can also connect base-level operators with one another to increase communication of problems and scale solutions.

Further development and maturation of solution concept: Through Agility Prime, AFWERX has been working to develop eVTOL technologies and support that industry. Through SBIR/STTR, AFWERX funds dual-use technology solutions developed by start-ups and refines them for end users.

S&T activities (includes experimentation): AFWERX facilitates S&T activities through its management of SBIR/STTR programs that fund early-stage research and development of technologies. AFWERX leverages DAF and DoD resources (including test infrastructure, certification authorities, interagency relationships, and early operational use cases) to support prototype demonstration and further experimentation.

Technology maturation (early stage): AFWERX supports early-stage maturation through management of the SBIR/STTR programs, including transitioning contracts from phase I to phase II and hopefully to the transition phase.

Funding and investment: AFWERX provides funding to ventures through SBIR/STTR phase investments, using SBIR/STTR Open Topic solicitations to transform the USAF and USSF into early-stage investors to accelerate the development and adoption of dual-use technology and grow the number of companies working with DAF.

Coordination: AFWERX works with AFRL and DAF customers and has partnered with other DIOs (or at least funded the same ventures that others, like NSIN, have funded). In the past, it partnered with an accelerator to help start-up ventures learn the basics of pitching their idea so that they could apply to SBIR/STTR programs and be ready to partner with the military. (AFWERX now runs its own accelerator to perform these roles.)

Hard infrastructure to support development: AFWERX has leveraged DAF and DoD resources such as test ranges and certification authorities to support prototype demonstration.

CTP-related outputs	Key outputs include SBIR/STTR agreements and prototypes of systems.
Definition of success	No single definition seems to exist, but AFWERX acknowledges that "transition can happen in different contexts: commercialization contract, prime contract, being sub to a prime." The organization also tries to use its phase III/strategic financing in a way that ensures its earlier phase money is not spent on ventures that never scale.[t] AFWERX may track the following outputs: • funded research and projects • scalable solutions • projects transitioned or deployed • ventures able to work with DAF and the wider government • domestic market capable projects producing commercial products for the military.
CTP-related outcomes	Any measures of success or effectiveness that AFWERX uses were not shared with the team at the time of this writing. One interviewee told the team that "success varies depending on the program. Our [AFVentures] measure of success is to enable the broader government to have access to companies or capabilities while providing money to get after it."[u] Another interviewee noted that "AFWERX does great at getting money out the door to small business."[v]

Success stories include Wickr, a company that was acquired by a larger entity after it was able to grow; and ICON, a 3D printing company that produces housing.[w] Another success is Ditto, a small California-based company that makes software that allows for data synchronization and application operation in areas where connectivity may be unreliable. Ditto has received an OTA contract from AFWERX and a phase III contract worth up to $950 million.[x]

---

[a] U.S. Air Force Public Affairs, "Air Force Opens Doors to Universities, Small Businesses and Entrepreneurs to Boost Innovation," July 21, 2017.

[b] Interview 33, Interviewee 44, October 13, 2021.

[c] AFWERX, "About Us," webpage, undated b.

[d] U.S. Air Force, *DoD FY 2023 Budget Estimates, Air Force Justification Book*, April 2022, p. 199.

[e] Kristin Mosbrucker, "New Air Force Innovation Hub in Austin," *San Antonio Business Journal*, December 12, 2018; Air Force Research Laboratory Public Affairs, "SpaceWERX Launch Announced," December 8, 2020.

[f] AFWERX, "Spark," webpage, undated d.

[g] AFWERX, undated b.

[h] AFWERX, "AFVentures," webpage, undated a; Interview 36, Interviewee 48, October 21, 2021.

[i] Interview 36, Interviewee 48, October 21, 2021.

[j] AFWERX , "Spark," webpage, undated c; and AFWERX, "Spark Tank," webpage, undated d.

[k] Interview 33, Interviewee 44, October 13, 2021.

[l] Interview 45, Interviewee 60, November 16, 2021.

[m] Air Force Research Laboratory Public Affairs, 2020; Air Force Research Laboratory Public Affairs, "SpaceWERX Ready to Propel Space Innovation," August 25, 2021.

[n] Brandi Vincent, "Space Force's Innovation Hub Announces Solicitations in the Works," *Nextgov.com*, August 12, 2021.

[o] SpaceWERX, "About Us," webpage, undated.

[p] Interview 33, Interviewee 44, October 13, 2021.

[q] Interview 44, Interviewee 57, November 16, 2021.

[r] Interview 44, Interviewee 57, November 16, 2021.

[s] Interview 36, Interviewee 48, October 21, 2021.

[t] Interview 36, Interviewee 48, October 21, 2021.

[u] Interview 36, Interviewee 48, October 21, 2021.

[v] Interview 33, Interviewee 44, October 13, 2021.

[w] Wickr, "Wickr Selected as Only Secure Communication Platform for Strategic Expansion Initiative with U.S. Air Force," press release, April 9, 2020.

[x] Tom Temin, "Who Says Small Innovators Can't Get Big Federal Contracts," *Federal News Network*, May 10, 2022.

ARCWERX

Snapshot of organization	**Founded:** Established in Tucson, Arizona, in 2019.[a]
	Mission and focus: ARCWERX's mission is to enable Air Reserve Component (ARC) personnel, which includes the Air National Guard and Air Force Reserve, to solve ARC and broader Air Force operational problems by leveraging members' civilian expertise. That civilian or private-sector expertise is considered an innovation resource unique to the ARC. ARCWERX's three priorities are to (1) increase innovation capacity—the "ability to produce and exploit new products, services or processes over long periods of time"[b]—in the ARC; (2) gather ARC problem statements and accelerate the development of scalable solutions developed by ARC personnel to address those problems; and (3) increase partner nations' capabilities by leveraging the Air National Guard (ANG) State Partnership Program[c] innovation ecosystem.[d]

Organizational statistics: As of 2021, ARCWERX had five full-time personnel. ARCWERX has funded almost 800 projects over the last four years addressing ARC-identified problems; however, it is not clear how many of these projects represented new capabilities or innovative technologies that transitioned to the field, as opposed to new processes or concepts.[e]

How it operates: ARCWERX's three lines of effort are education, connection, and facilitation activities.[f] ARCWERX seeks to build innovation capacity through education and training so that ARC personnel can solve problems by executing locally developed projects to solve operational and strategic problems at the unit level. ARCWERX's unique approach is to leverage the civilian skills and expertise of ARC personnel as well as draw on local innovation resources such as universities, trade organizations, and commercial business located close to an ARC unit location.[g]

ARCWERX educates and trains ARC personnel in lean methods[h] and business practices, including wing-level training on building innovation cells and courses in design thinking and lean start-up, so that ARC personnel can execute locally developed projects. ARCWERX connects ARC units and personnel to the ARCWERX innovation ecosystem, which includes industry, academia, research organizations, and other DIOs, to find the "right resource at the right time" for customers' projects.[i] To facilitate this problem-solution matching function, ARCWERX is developing a data set that stores problems, innovative ideas, and solutions, as well as a larger network of contacts throughout industry, academia, and defense innovation organizations.

Primary customers: Air Force Reserve (AFR), Air National Guard (ANG), and U.S. Air Force.

Link to CTP

Core functions performed: Technology scouting; problem identification, validation, and curation; network development; solution concept generation; matching problems to solutions; technology maturation (early stage); further development and maturation of solution concept; RDT&E activities; transition to fielding (limited).

Enabling functions performed: Funding, coordination, navigation and relationships, information sharing and reporting.

How they do their CTP functions

Technology scouting: ARCWERX performs outreach and discovery activities to find commercial companies and technologies (including but not limited to those associated with ARC members) and maintains a database of innovation ecosystem partners, programs, and companies and technologies that match to ARC problems. ARCWERX also reviews the SBIR database (SBIR.gov) as well as ventures coming out of AFWERX-sponsored accelerator programs.[j] ARCWERX activities are not limited to ARC personnel and associated commercial businesses. ARCWERX often has active-duty participants, and they also include many academic entities in their activities.

Problem identification, validation, and curation: ARCWERX solicits statements of operational problems from ARC personnel, who then submit problems on ARCWERX's website. Problem identification starts at the ARC unit level, but it looks for opportunities to scale beyond that initial unit to additional units and the active Air Force. Two full-time innovation consultants (uniformed personnel) evaluate the feasibility and impact of addressing these problems, seeking to identify project ideas that have the greatest impact across the force (i.e., save lives, time, or money and scale beyond the base or wing level where possible) and that are affordable within ARCWERX's portfolio.[k]

ARCWERX receives approximately 300 problem statements per year from ARC and down-selects, based on funding, to roughly 150 to work on in each year.[l]

Solution concept generation: ARCWERX reaches out and encourages ARC personnel to leverage their civilian experience and knowledge to development and submit potential solution ideas for ARC problems.

ARCWERX occasionally seeks out airmen with specific nonmilitary skills to leverage civilian experience in support of projects. ARCWERX also provides education that encourages innovation leads at the wing level to make use of the civilian expertise of their members. Additionally, there are multiple ongoing projects to catalog the civilian employment experience of willing ARC members.

Network development: ARCWERX identifies and connects ARC personnel and other innovation partners through regular in-person and virtual events, seeking to develop a network of ARC personnel to share innovative ideas and technology.[m] "We build on the unique structure of the Air Guard and Reserve, which does not exist in the active component. There are very strong ties to the local university and business ecosystem, but they don't know how to utilize it. We maintain a Rolodex of all the ecosystem partners, all the programs, and a growing library of interesting technologies and companies they can connect to."[n] ARCWERX is developing an Innovation Portal to facilitate communication and the sharing of ideas across the enterprise.[o]

ARCWERX identifies and curates problem statements for NSIN's Pacific-Southwest regional team and tries to match these problems to applicable NSIN programs.

Matching problems to solutions: ARCWERX's original approach started with identifying a solution—a commercial technology—and then searching for a military problem it could address. ARCWERX concluded that this approach did not work well and revised its approach to start with the military problem instead: "We start with a problem, run it through the pipeline, and continue funding based on progress to move into transition. It's about reducing uncertainty. Every step, we ask if people are willing to use it, if it's feasible in the budget, [and] if we are able to produce results."[p]

After ARCWERX selects a problem, it conducts a workshop with ARCWERX personnel to help the submitting unit understand the core problem, stakeholders, customers, assumptions, and test alternatives. Workshops help the team (submitting unit and ARCWERX) determine if the problem is worth solving, if potential solutions and necessary resources are available, and whether a prototype is worth those resources. This process includes evaluating what other DIOs are working on and whether ARCWERX can leverage other projects before committing resources.

Each year, ARCWERX selects approximately 150 projects for one of two paths. Approximately 120 projects aim to use mature technology to address what is believed to be an understood problem. In addition, approximately 30 problems that are not fully understood or lack a technologically mature solution are also selected.

Further maturation of solution concept: The ARCWERX incubator, introduced in FY 2022, includes activities to further mature the solution concept. Projects are taken from idea through facilitation to proof-of-concept. If the proof-of-concept demonstration is successful, a pilot program is sponsored at multiple locations. During the pilot phase, the project is assessed for acceptance and sustainability prior to transition.

Technology maturation (early stage): ARCWERX seeks to support technology maturation. One line of effort—facilitation—is intended to "create a system where innovative ideas can be shared, tested and scaled across the enterprise."[q] ARCWERX plans to introduce an incubator program that includes limited funding to further develop the solution concept and build and test a prototype to demonstrate that the solution works.[r]

RDT&E activities: ARCWERX leverages the flexibility of ARC personnel to work on specific projects. Because of fiscal constraints, ARCWERX limits funding of RDT&E efforts to the provision of matching funds for AFWERX phase II SBIR submissions. All ARCWERX funding is O&M.

Transition: Throughout a project, ARCWERX seeks to reduce technical uncertainty, asking if units will be willing to use the prototype and if it is feasible based on the budget. Beginning at the pilot stage, transition advocates are brought in to facilitate transition planning and budgeting; ARCWERX's goal is to transition approximately 10% of facilitated projects, or one to three projects per year.[s]

ARCWERX defines transition in two ways: (1) An Air Force or ARC program office takes responsibility for future development and funding of a program, or (2) a sister service sees promise in the program and agrees to take responsibility for future development or funding. In the latter case, transition often comes at an early stage of product maturity.

Funding: ARCWERX provides funding to support development based on progress and results. Total funding was $9.2 million in 2018 and 2019, $6.9 million in 2020, $5.5 million in 2021, and an expected $8.3 million in 2022.[t] ARCWERX funding is limited to ARC personnel and priorities.

Coordination: ARCWERX states that it tracks what all the units in the Air Guard and Reserve are working on, so it can connect units that have similar problems to potentially achieve economies of scale when building prototypes, share best practices, and pair with functional expertise.[u]

Navigation and relationships: ARCWERX helps a commercial business navigate DoD to find potential customers.[v]

Information sharing and reporting: ARCWERX is developing an Innovation Portal to facilitate information sharing across the enterprise.

ARCWERX uses Air Force platforms to manage the flow of information across the enterprise. ARCWERX receives problem statements and runs "campaigns" on the Guardian and Airmen Innovation Network (GAIN). Once an idea or proposal becomes an active project, it manages those projects on the VISION portal to share it with the rest of the Air Force.

CTP-related outputs	ARCWERX defines output in two ways: (1) applications for new projects; and (2) a proposal's potential impact, categorized as either revolutionary, transformative, or incremental.
Definition of success	ARCWERX defines success as adding value for ARC leadership. ARCWERX reports output using return on investment (ROI). It queries project managers for their assessment of man-hours and dollars saved as a result of their project, then aggregates those numbers. ARCWERX then uses the cost of one enlisted man-day ($250) to convert time to dollars saved.
CTP-related outcomes	ARCWERX measures outcomes through the solving of customer problems. That can be accomplished through funding, providing connections to resources, finding a commercial technology, or helping with designing in-house solutions.

[a] LinkedIn, "ARCWERX," undated.

[b] David Wilkinson, "Innovation Capacity: How to Develop It in Your Organization," *The OR Briefings,* blog, undated.

[c] The State Partnership Program pairs a state's National Guard with a partner country to create a mutually beneficial relationship in which partners conduct military-to-military engagements and leverage "whole-of-society relationships and capabilities to facilitate broader interagency and corollary engagements spanning military, government, economic and social spheres." National Guard, "State Partnership Program," webpage, undated.

[d] DODX information packet obtained by authors through communication with AAL.

[e] Interview 9, Interviewee 10, August 18, 2021; ARCWERX overview brief, chart 8, provided to authors, August 18, 2021.

[f] ARCWERX information to authors, August 18, 2021.

[g] Interview 9, Interviewee 10, August 18, 2021.

[h] Lean methods are a set of principles and practices to improve efficiency and effectiveness while reducing waste and using fewer resources. For a nonmilitary but useful explanation, see Sharon A. Schweikhart and Allard E. Dembe, "The Applicability of Lean and Six Sigma Techniques to Clinical and Translational Research," *Journal of Investigative Medicine,* Vol. 57, No. 7, October 2009, pp. 748–755.

[i] Interview 9, Interviewee 10, August 18, 2021.

[j] Interview 9, Interviewee 10, August 18, 2021.

[k] ARCWERX information to authors, August 18, 2021.

[l] ARCWERX information to authors, August 18, 2021.

[m] ARCWERX information to authors, August 18, 2021.

[n] Interview 9, Interviewee 10, August 18, 2021.

[o] ARCWERX information to authors, August 18, 2021.

[p] Interview 9, Interviewee 10, August 18, 2021.

[q] ARCWERX information to authors, August 18, 2021.

[r] ARCWERX information to authors, August 18, 2021.

[s] ARCWERX overview brief, chart 9 and chart 8, provided to authors, August 18, 2021.

[t] ARCWERX overview brief, chart 8, provided to authors, August 18, 2021.

[u] Jennifer-Leigh Oprihory, "ARCWERX Will Leverage ANG, Reserve Innovation to Enhance Total Force," *Air & Space Forces Magazine*, May 13, 2020.

[v] Interview 9, Interviewee 10, August 18, 2021.

Capability Prototyping/Rapid Reaction Technology Office

Snapshot of organization	**Founded:** Originally founded in 2001 as the Rapid Reaction Technology Office, this organization was renamed Capability Prototyping in 2022. **Mission and focus:** CP/RRTO's mission is to identify, support development, and facilitate adoption of new innovative capabilities by the Joint Force.[a] CP/RRTO aims to create faster pathways for concepts to become capabilities by identifying promising emerging technologies and capabilities, supporting prototyping and experimentation to mature them, and transitioning them into programs.[b] **Organizational statistics:** In FY 2021, CP/RRTO had a budget of $224 million that was spread across three prototyping programs,[c] in addition to $5 million for four demonstration/experimentation programs. There are $264 million in active projects underway in this fourth program, with transitions beginning in FY 2022.[d] **How it operates:** CP/RRTO funds and facilitates the development of prototypes, demonstrations, and experimentation with promising technologies that address National Defense Strategy and OSD/R&E modernization priorities; it also helps government sponsors identify emerging and nontraditional businesses with promising solutions to their needs.[e] **Primary customers:** Joint Staff, CCMDs, services, DoD agencies, and joint warfighter.[f]
Link to CTP	**Identification functions performed:** Technology scouting; customer discovery; problem identification, validation, and curation; matching problems to potential solutions. **Development functions performed:** S&T activities (experimentation), RDT&E activities (prototyping, demonstrations). **Adoption functions performed:** Limited fielding, fielding at scale. **Enabling functions performed:** Funding and investment, hard infrastructure to support development (test platforms/equipment), navigation and relationships (facilitating business entry to DoD and interactions with DoD, mentoring).
How they do their CTP functions	**Technology scouting:** • CP/RRTO seeks to identify "innovative, leap-ahead capabilities" from all sources that solve problems tied to National Defense Strategy modernization priorities and serve joint objectives for which CP/RRTO can secure "co-funding from other stakeholders" and identify a "clear transition path to programs or people."[g]

- To identify technologies, CP/RRTO uses two streamlined proposal processes. Their principal means of identifying technologies is through a white-paper submission process: CP/RRTO accepts proposals on a rolling basis through the year, welcoming three-page white papers from interested parties with ideas. Ideas may be submitted by anyone, including small businesses, traditional and nontraditional DoD contractors, academia, CCMDs, military services, government labs, FFRDCs, and UARCs. Each proposal is screened through the lens of DoD's strategic priorities and CP/RRTO's seven guiding principles for allocating funding.[h] To supplement and complement its white-paper review process, CP/RRTO also requests 100-word submissions from innovative performers through SAM.gov. (These efforts and the screening process are described below.)

Technology scouting; matching problems to potential solutions:

- CP/RRTO's primary intake method is through submission of a white paper. Interested parties with ideas submit a short application including a three-page white paper that specifies what their technology can do.[i] On a quarterly basis, CP/RRTO convenes technical and operational SMEs, drawing on representatives from more than 100 entities including CCMDs, MAJCOMs, defense innovation organizations, and others, to review the concepts and provide recommendations on proposals they have received.[j] Based on this review, companies that have the most compelling solutions as determined by CP/RRTO have the potential to be selected for pilot prototyping, demonstration, or experimentation opportunities, described in greater detail below.[k] CP/RRTO seeks to grant awards throughout the year and in the same year of execution, promising a rapid turnaround time.[l]

- As CP/RRTO evaluates technologies, it determines which DoD organizations the technology might fit into. CP/RRTO sends out the company's pitch deck to prospective partners to see if the technology is of interest. If the end user and/or program determine the technology may be useful, CP/RRTO sets up a meeting with the company. Based on that conversation, CP/RRTO may set up a government-only meeting with the prospective partner to discuss a path to prototyping the technology, whether the end user or program would potentially be able to integrate the technology into their technology spiral, and whether the end user or program would cofinanced. If the government sponsor has money and a requirement, CP/RRTO will bring the technology into a prototyping program; for example, if the sponsor can use PEO funding to bridge the first one to two years, they can seek to have funds added to the program objective memorandum for later years. CP/RRTO will sometimes support development even without a transition partner, particularly when the technology is cross-cutting and could be used by many DoD end users.[m]

Technology scouting; customer discovery; problem identification, validation, and curation; matching problems to potential solutions: CP/RRTO also conducts targeted innovation outreach efforts to supplement and complement its white-paper review process. CP/RRTO conducts innovation outreach typically on behalf of a government sponsor, targeting emerging and nontraditional technology companies. CP/RRTO convenes a "needs meeting" to develop needs, bringing together a government sponsor and technology experts, including reps from venture capital businesses. CP/RRTO then summarizes and publishes those needs on the SAM.gov website three to four times per year. Companies are asked to make a 100-word submission to respond to the SAM.gov special notice. The government sponsor selects companies, and CP/RRTO convenes a solutions meeting where companies brief the government sponsor on their capabilities and solutions. The government sponsor then identifies solutions of interest and can fund experimentation or demonstration, conduct experiments, or procure successfully demonstrated products.[n]

Customer discovery: If CP/RRTO finds a technology or prototype that could be useful to a potential DoD customer, it serves as a liaison between the technology and potential customer. CP/RRTO provides the potential DoD customer with information about the technology and asks them if they are interested in maturing it further. If the customer says yes, CP/RRTO will work with the DoD customer and attempt to develop an executable program with defined deliverables, metrics, and a transition plan. If the idea is not a good fit for CP/RRTO (e.g., the idea is already too mature for prototyping), then CP/RRTO simply facilitates the introduction.

Development functions: CP/RRTO supports the development of promising technologies in one of two ways: providing funding to develop prototypes, and providing access to demonstration and experimentation venues.

S&T activities (experimentation) and RDT&E activities (prototyping, demonstrations); funding and investment

Funding to develop prototypes: The following information comes from an RRTO/CP document from January 2022, on file with authors, unless otherwise indicated. CP/RRTO has three primary funding lines that support development of prototypes (a fourth funding line was not included in the FY 2021 budget).

- Emerging Capabilities Technology Development: Through the program, CP/RRTO provides funding to develop "proof-of-principle" prototypes of emerging technologies and prototypes to "explore the art of the possible." Projects tend to be one to three years long and less than $6 million. Intake process: accepts walk-ins and emails throughout the year.
- Quick Reaction Special Projects (QRSP): Through this program, CP/RRTO develops prototypes with innovative performers that address immediate needs and focus on delivering "quick wins for [the] joint warfighter." Projects average 12 months and less than $1 million. QRSP is focused on nontraditional solutions, risk-taking, and innovation.[o] Intake process: accepts walk-ins and emails throughout the year.
- Rapid Prototyping Program: Through this program, CP/RRTO funds and develops "operational prototypes that deliver Joint Modernization capabilities" and seeks to help move these "technologies across [the] 'valley of death' and into Service programs of record."[p] Projects average less than $15 million across one year.[q] Intake process: issues a call for proposals to the military services and defense agencies. (As of April 2022, this Rapid Prototyping Program was transferred to another office.)

Demonstration and experimentation; funding and investment; hard infrastructure to support development: CP/RRTO provides funding for three demonstration venues—High Speed, Electronic Keel Marine Testbed (Stiletto); Multi-Intelligence and ISR Technology Demonstration Venue (Thunderstorm); and Joint Interagency Field Experimentation (JIFX)—and it makes these venues available to companies with promising technologies.[r] The results from these experiments and demonstrations are made available to the government.[s]

Transition to fielding (limited fielding and fielding at scale): CP/RRTO provides support to transition technologies to DoD for limited fielding and fielding at scale:

- CP/RRTO selects TRL 3 to 4 technologies that have a high likelihood of transition, and fosters and supports relationships and partnerships with transition partners (programs and end users) to increase adoption rates of technologies that CP/RRTO has supported. Together, those things set conditions for limited fielding and fielding at scale.
  - CP/RRTO builds partnerships with end users and programs at the outset of the prototyping process, to facilitate later adoption and fielding.
  - CP/RRTO supports development of technologies where the government transition partner and company are already connected. CP/RRTO supports some companies that have already identified government sponsor–transition partners but need additional financial support to develop a prototype. CP/RRTO supports some government sponsors that have already identified a company with promising technology and solved the programmatic issues but need additional financial support to develop a prototype.

Funding and investment: CP/RRTO provides funding and investment to support the development of promising concepts and technologies. Among other guiding principles, CP/RRTO seeks to "invest where others do not," filling "seams, cracks, and fissures," to secure "co-funding from other stakeholders," and to select technologies with a "clear transfer/transition path."[t]

Hard infrastructure to support development: CP/RRTO facilitates demonstrations and experiments for prototypes, providing venues and support infrastructure to enable concept of operations for prototypes.[u] These venues, such as Thunderstorm, make it easier for nontraditional and small businesses to bring their ideas and prototypes to the DoD.[v]

Navigation and relationships: CP/RRTO serves as a liaison between DoD customers and technology companies, actively engaging with both emerging and nontraditional companies to quickly bring capabilities into DoD.[w] CP/RRTO coaches and mentors businesses, providing guidance and expectations about warfighter needs and environment while developing their technologies.[x]

CTP-related outputs	The outputs as described here are taken largely from the RRTO/CP document from January 2022, on file with the authors, unless otherwise indicated:
	
	• Creation of accelerated pathways with minimal bureaucratic hurdles (low barrier to entry, quick decision time) for ideas to become joint capabilities.
	• Identification of promising technologies.
	• Identification of nontraditional solution providers.
	• Identification and validation of DoD problems.
	• Analysis of promising technologies.
	• Prototypes of promising technologies built (see outcomes section below for an example).
	• Demonstrations of promising technologies conducted: In FY 2021, 219 technologies were demonstrated to a total of 715 DoD/government attendees.[y]
	• Experiments of promising technologies conducted.
	• Mentoring of small businesses: In FY 2021, 136 small businesses were mentored.[z]
	• Education and support of government sponsors to help them engage with emerging and nontraditional technology companies.
	• Engagements between companies and potential government sponsors (customers and end users) through innovation outreach program. For example, in FY21, 1,070 emerging and nontraditional technology companies applied to participate in CP/RRTO's innovation outreach program; 95 companies presented to the requesting government audience, and 24 companies participated in follow-on discussions/negotiations.
	• Transition of concepts into capabilities. Specific outputs related to transition in FY 2020 were identified:[aa]
	  – Fielding at scale and limited fielding of technologies: CP/RRTO transitioned 36 (60%) technologies from CP/RRTO prototyping programs to "capability delivery," meaning that the prototypes "were delivered to program of record or joint warfighter as a capability, or significant component of a capability." More specifically, CP/RRTO defines this transition success as the "delivery of capability to the service, CCMD, or warfighter."
	  – Information to inform future capability decisions: CP/RRTO "transitioned" 19 (32%) technologies from CP/RRTO prototyping programs to "capability enabler," meaning that the prototypes or analyses "explore emerging technologies to inform capability decisions without committing major resources." CP/RRTO measures "transition success for capability enabler" by "value added when informing future capability decisions.
	
	CP/RRTO did not transition five (8%) technologies from CP/RRTO prototyping programs. "No transition" is defined as "any project that does not meet criteria for transition as capability delivery or capability enabler."

Definition of success	CP/RRTO strives for a balance between having a high transition rate and allowing its funding programs to pursue riskier things—a 100% transition rate signifies that they are being too safe and risk averse, so they set a goal of successfully transferring 80% of their prototypes.[bb] CP/RRTO recognizes two different types of transition: capability delivery (i.e., a prototype that is delivered to a program of record or joint warfighter as a capability) and capability enabler (i.e., a prototype or analysis that informs capability decisions without committing major resources).[cc]

CTP-related outcomes	CP/RRTO identifies, matures, tests, and facilitates the fielding and adoption of innovative technologies by the Joint Force, thereby increasing warfighter readiness and effectiveness.
	As an example, CP/RRTO provided $2.037 million in prototype funding to Maritime Applied Physics Corporation (MAPC) for automatically deploying and retrieving a parafoil system with its large jet ski platform, the Greenough Advanced Rescue Craft (GARC). This concept system allows for greater digital communications and sensor connectivity. CP/RRTO, recognizing the potential for payoff, invested early to develop this capability and mature the idea so it could eventually be demonstrated for its feasibility. Because of CP/RRTO's support, MAPC successfully transitioned to the Navy's PEO, which identified this technology as a solution for a program of record. CP/RRTO's funding was key to finishing RDT&E efforts, eventual transition, maturation, and production under the program of record.[dd]

[a] RRTO/CP document, on file with authors, January 2022.

[b] RRTO, "Overview," slide presentation, February 2021.

[c] RRTO/CP document, on file with authors, January 2022.

[d] RRTO information to authors, April 27, 2022.

[e] RRTO, 2021.

[f] RRTO/CP document, on file with authors, January 2022.

[g] RRTO/CP, document, on file with authors, January 2022.

[h] RRTO, "Overview," press release, January 2018.

[i] CP/RRTO information to authors, April 27, 2022; RRTO, 2021; RRTO/CP document on file with authors, January 2022.

[j] RRTO, 2018.

[k] RRTO, "2021 Global Needs Statement," SAM.gov, March 24, 2021.

[l] DAU, "Small Business Innovation," May 19, 2021.

[m] CP/RRTO information to authors, April 27, 2022.

[n] RRTO/CP, document on file with authors, January 2022.

[o] Interview 28, interviewee 35, September 28, 2021.

[p] RRTO/CP, document on file with authors, January 2022; RRTO, 2021.

[q] RRTO/CP, document on file with authors, January 2022.

[r] RRTO, 2021.

[s] CP/RRTO information to authors, April 27, 2022.

[t] RRTO, 2021.

[u] RRTO, 2021.

[v] Interview 28, Interviewee 35, September 28, 2021.

[w] RRTO/CP, document on file with authors, January 2022.

[x] Interview 28, Interviewee 35, September 28, 2021.

[y] RRTO/CP, document on file with authors, January 2022.

[z] RRTO/CP, document on file with authors, January 2022.

[aa] The specific outputs listed for FY 2020 are from RRTO/CP, "Transition Summit Excerpt," on file with authors, undated.

[bb] Interview 28, interviewee 35, September 28, 2021.

[cc] RRTO/CP, "Transition Summit Excerpt," on file with authors, undated.

[dd] CP/RRTO information to authors, April 27, 2022; see also Xavier Vavasseur, "U.S. Navy Successfully Tests GARC/TALONS for LCS MCM Mission Package," *NavalNews.com*, November 26, 2019.

DARPA (Defense Advanced Research Projects Agency)

Snapshot of organization	**Founded:** DARPA was created in 1958 in the wake of the Sputnik launch.
	Mission and focus: DARPA is an independent R&D agency within the DoD. DARPA aims to maintain U.S. technological superiority over adversaries by selecting high-risk, high-payoff, and the most forward-looking technologies that might have the largest impact on future national security challenges. DARPA seeks to "make pivotal investments in breakthrough technologies for national security."[a]
	Organizational statistics: DARPA's annual budget is $3.5 billion (FY 2021 enacted). It has a staff of 220 government employees in six technical offices and nearly 100 program managers overseeing about 250 programs.[b]
	How it operates: DARPA projects are driven by PMs and focus on four strategic imperatives: defend the homeland, deter and prevail against high-end adversaries, prosecute stabilization efforts, and advance foundational research in science and technology.[c] The agency focuses on early-stage ideas and research that has potential for high-payoff capabilities but entails risk due to the uncertainty surrounding the time frame or likelihood of successful development. Once a project is approved by the DARPA director, the agency partners with defense companies, small and large commercial businesses, start-ups, allied nations, and academia to demonstrate proofs-of-concept for technologies applicable to a military need or future military application.
	Primary customers: DoD-wide.
Link to CTP	**Identification functions performed:** Technology scouting; problem identification, validation, and curation; solution concept generation; and matching problems to potential solutions.
	Development functions performed: S&T activities, technology maturation (early stage), product maturation (early stage), product maturation (late stage).
	Adoption functions performed: Transition to fielding (limited fielding, fielding at scale).
	Enabling functions performed: Funding and investment, coordination, navigation and relationships, hard infrastructure to support development.
How they do their CTP functions	**Technology scouting:** DARPA's technical PMs are responsible for maintaining awareness of technology trends and capitalizing on them before others do.[d] PMs typically serve three- to five-year terms and are drawn from across industry and research institutions. PMs pitch potential projects to the DARPA director and deputy director. Using their previous industry experience, PMs choose potential solutions that address challenges, from deep science to systems to capabilities.[e] DARPA also posts requests for information (RFIs) on SAM.gov and Grants.gov to learn about emerging technologies. Given DARPA's long-standing reputation for pursuing innovative ideas, researchers, entrepreneurs, and government personnel also contact PMs directly to discuss research or project ideas, emerging technologies, and technology transition and can attend "proposer days," where PMs discuss new or upcoming programs.[f]
	Problem identification, validation, and curation: DARPA can source problems from any DoD entity or based on PM ideas for new programs that will address national security challenges. DARPA programs do not always address stated problems. They often seek to pursue opportunities to create new technology or capability not previously envisioned. DARPA has liaisons from the military services, intelligence community, and special forces whose responsibilities are to help align and tailor DARPA programs and research with military service problems and requirements when possible; however, DARPA does not require a specific military service sponsor or DoD customer to create a program. DARPA projects have some military use in mind when conceived, but the services may not be ready for a DARPA product initially, according to one interviewee, giving the example of the Air Force and early development of unmanned aerial vehicles. DARPA also has projects in areas that can have both defense-related and broader commercial applications such as mRNA (messenger ribonucleic acid) vaccines.[g]

Solution concept generation: DARPA often uses a two-step process when soliciting proposals. In the first phase, DARPA requests an abstract or a white paper from interested parties outlining areas such as proposed technical solution, risk, budget, schedule, and teaming arrangements. DARPA provides comments and feedback to interested parties on those abstracts and an "encourage" or "discourage" letter recommending that parties submit a full proposal in response to a DARPA solicitation; however, this preliminary assessment does not preclude parties from submitting a proposal.[h] In addition to solicitations for individual programs, each technical office also issues an "office-wide" broad agency announcement (BAA) to solicit ideas that may not align with a specific program but could have national security applications.[i]

Matching problems to potential solutions: DARPA publishes information about its program portfolio and technology offices.[j] Additionally, DARPA posts BAAs, requests for proposals (RFPs), and RFIs to advertise opportunities, solicit proposals, and learn from industry. Problems and potential solutions can come from any source but must have a national security application. It is the PM's responsibility to match a potential solution to the problem area of interest.

S&T activities: Nearly all of DARPA's budget is for basic research, applied research, and advanced technology development (6.1 to 6.3 RDT&E budget activity categories). Dr. Steven Walker, then DARPA director, described DARPA's role as "to do the fundamental research, the proof of principle, and the early stages of technology development that take impossible ideas to the point of implausible, but surprisingly possible."[k] DARPA typically does this work through contract vehicles and agreements, including FAR-based contracts, BAAs, SBIRs, OTAs, grants, CRADAs, and prize challenges with industry and universities.[l]

Technology maturation (early stage): Each DARPA program has specific metrics and milestones for each phase of the program. Larger programs typically last three to four years with multiple performers. Each phase may start with four to five performer teams with DARPA down-selecting based on performance during the phase. DARPA also uses SBIR phase I and II awards.

Product maturation (late stage): DARPA partners with the services and DoD components to conduct demonstrations of new technologies.[m] DARPA works to integrate existing and emerging technologies to demonstrate technical and operational capability in realistic test environments.[n]

Transition to fielding (limited fielding and fielding at scale): "DARPA typically stops [providing funding to support technologies] after demonstrating [their] feasibility," but it provides other forms of support to transition technologies to DoD or the commercial market.[o] DARPA can work with the military services or other DoD components to establish joint development offices to oversee a program and support the transition of early operational capability to the warfighter. An example is the long-range anti-ship missile (LRASM), which started as a program in 2009 and provided early operational capability in 2018.[p]

DARPA's Small Business Programs Office has a Transition and Commercialization Support Program (TCSP), intended to maximize the potential for companies to move their technologies into other R&D programs for further maturity.[q] These can include solutions or products for DoD acquisition programs, other federal programs, or the commercial market. TCSP is offered to support both phase I and phase II DARPA-funded SBIR/STTR projects.

Funding and investment: DARPA takes a portfolio approach to investments and activities by addressing a wide of range of challenges and opportunities at the same time. DARPA uses metrics and milestones to measure progress and ends projects if they are not meeting those.[r] DARPA uses the full range of contracting vehicles (e.g., FAR-based, OTA, SBIRs, CRADAs, prize challenges.) It seeks solutions for problems that are not requirements based but rather are centered around capability, needs, or opportunities.

Coordination: DARPA collaborates and partners with a wide range of stakeholders, including universities, industry, small businesses, government, the public, and the media.

Navigation and relationships: DARPA maintains relationships with numerous academic institutions, research labs, the military services, other government agencies, start-ups and other commercial companies, and allied nations. DARPA's liaisons help sustain relationships with the services, special operations community, and the intelligence community. DARPA has created the Embedded Entrepreneur Initiative (EEI) to help DARPA performers create businesses that can turn DARPA innovations and technologies into products for the commercial and defense markets.[s]

The EEI program embeds entrepreneurs into selected small businesses that are already on contract to DARPA. These entrepreneurs augment the scientific research teams with business expertise. "The seasoned entrepreneur or business executive [is funded with non-dilutive funding] for one to two years with the goal of developing a go-to-market strategy for both defense and commercial markets."[t] EEI also provides exposure to venture capitalists in DARPA-hosted events to provide a venue for early-stage ventures to hone their presentations and potentially garner commercial investment, ensuring vital emerging technology develops domestically and does not lead to problematic overseas investment.

Hard infrastructure to support development: Although DARPA does not directly perform research activities or operate its own laboratories, it can fund the use of hard infrastructure such as test ranges. DARPA often concludes memorandums of agreement with service partners to transition technology, which can include the services providing test and evaluation, range time, and operational personnel to support field exercises and demonstrations. DARPA performers carry out their R&D programs through contracts with their collaborators: industry, universities, nonprofit organizations, and federal laboratories.[u]

CTP-related outputs	"Everything that DARPA does has a military objective." DARPA seeks to produce (or allow its partners to produce) transformative technologies, primarily for military and national security use. DARPA also seeks to identify commercial uses for these technologies and to support the transition of these technologies to the commercial sector, which in many cases is essential to the viability of the new-entrant business. While "DARPA never does anything just for commercial benefit," commercialization can "drive the volume up, which drives the reliability up, which drives the availability, which drives down costs."[v]
Definition of success	DARPA's definition of success is to enable the development of breakthrough technologies for national security purposes. DARPA invests in high-risk areas that promise high reward, "transforming revolutionary concepts and seeming impossibilities into practical capabilities."[w]
CTP-related outcomes	DARPA defines innovation as "developing a unique technology that actually gets used. If it doesn't get used, it shouldn't be considered innovation. Transition is therefore important—and how it gets transitioned."[x] A DARPA program can transition from research to impact in a wide variety of ways. Some programs may result in technology that has immediate military use while other technologies may transition into other development programs, into the commercial space for further development and maturation before being incorporated into military systems, or into a program that supports another government agency's mission.

According to interviewee comments and DARPA literature,[y] DARPA defines eight modes of transition where technology can be

- incorporated in a program of record
- used to establish a standard in a community
- transitioned to commercial market
- further developed by a DoD component (service)
- further developed by others
- used to enable the mission of another government agency
- inserted into an ongoing or future DARPA program
- transitioned directly to operational use.

Technologies may transition in multiple ways. Not all DARPA efforts transition successfully, but they can serve as learning experiences. Because of DARPA's culture of risk-taking for innovation, some programs do not achieve all their objectives or do not transition directly.[z]

- Examples of military success: LRASM, precision-guided munitions, stealth technology, UAVs, infrared night vision.
- Examples of dual-use technologies: miniature GPS receivers, automated voice recognition, personal electronics.
- Examples of technology that did not transition directly but served as a learning experience: The Falcon Hypersonic Technology Vehicle 2 did not achieve its goal of sustained, controlled Mach 20 glide flight, but it produced insights, data, and technology advances that informed later DARPA and service hypersonic programs such as the Hypersonic Air-breathing Weapon Concept.[aa] DARPA conducted two flight tests in 2010 and 2011, each of which did not complete the full planned test time of 30 minutes.[bb]

---

[a] DARPA, "Breakthrough Technologies for National Security," March 2015.

[b] DARPA, "What DARPA Does," webpage, undated i.

[c] DARPA, "Creating Technology Breakthroughs and New Capabilities for National Security," August 2019.

[d] DARPA, 2019.

[e] DARPA, "About DARPA," webpage, undated a.

[f] DARPA, "Industry," webpage, undated f.

[g] Interview 14, Interviewee 16, September 2, 2021.

[h] Interview 14, Interviewee 16, September 2, 2021.

[i] DARPA, "Office-wide Broad Agency Announcements," website, undated g.

[j] DARPA, "Our Research," webpage, undated h.

[k] Marcy E. Gallo, *Defense Advanced Research Projects Agency: Overview and Issues for Congress*, Washington, D.C.: Congressional Research Service, R-45088, August 19, 2021.

[l] Interview 13, Interviewee 15, August 27, 2021.

[m] DARPA, 2019.

[n] DARPA, 2019, p. 26.

[o] Interview 53, interviewee 71, June 17, 2022.

[p] DARPA, 2019, p. 17.

[q] DARPA, "Small Business Programs Office Transition and Commercialization Support Program Fact Sheet," May 21, 2020.

[r] DARPA, 2020.

[s] DARPA, "Embedded Entrepreneurship Initiative," webpage, undated c.

[t] Barbara McQuiston, "Statement on Performing the Duties of the Under Secretary of Defense for Research and Engineering," presented to the U.S. Senate, Subcommittee of the Committee on Appropriations, Washington, D.C. April 13, 2021.

[u] Gallo, 2021.

[v] Interview 53, interviewee 71, June 17, 2022.

[w] Interview 53, Interviewee 71, June 17, 2022; see also Stefanie Tompkins, "Statement by Dr. Stefanie Tompkins, Director Defense Advanced Research Projects Agency (DARPA)," submitted to the U.S. Senate Armed Services Committee, Subcommittee on Emerging Threats and Capabilities, April 6, 2022.

[x] Interview 53, interviewee 71, June 17, 2022.

[y] Interview 53, interviewee 71, June 17, 2022; DARPA, 2019.

[z] DARPA, 2019.

[aa] Interview 53, Interviewee 71, June 17, 2022; DARPA, "Falcon HTV-2 (Archived)," website, undated d; DARPA, "Hypersonic Air-breathing Weapon Concept (HAWC)," website, undated e.

[bb] Space.com, "Superfast Military Aircraft Hit Mach 20 Before Ocean Crash, DARPA Says," August 18, 2011.

Defense Innovation Unit

Snapshot of organization	**Founded:** Headquartered in Mountain View, California, DIU was established in 2015. **Mission and focus:** DIU "strengthens our national security by accelerating the adoption of leading commercial technology throughout the military and growing the national security innovation base." DIU seeks to "transform military capacity and capabilities" by increasing the rate and use of commercially derived technologies in military systems.[a] **Organizational statistics:** Core DIU operated a $50 million RDT&E budget in FY 2021, an increase from $10 million in FY 2017.[b] DIU's staff is a mix of civilian and military personnel: "As of September 2021, DIU's staff included 23 civilians, 10 special government employees, 34 active-duty military personnel, 38 part-time reservists, three Intergovernmental Personnel Act staffers, 26 detailees and liaisons, and 77 full- and part-time contractors providing flexible, specialty expertise."[c] **How it operates:** DIU serves as a liaison between DoD entities and commercial technology companies by soliciting proposals and contracting with companies that have innovative solutions that meet the needs of DoD partners. DIU operates five geographically dispersed offices in Mountain View, Boston, Austin, Chicago, and the Pentagon in Arlington, Virginia, to connect with local entrepreneurs. DIU also exercises oversight over the National Security Innovation Capital (NSIC) initiative and the NSIN, which operate as independent and complementary organizations.[d] **Primary customers:** DoD services, CCMDs, defense agencies, and other components.[e]
Link to CTP	**Identification functions performed:** Technology scouting; network development; problem identification, validation, and curation; solution concept generation; matching problems to potential solutions. **Development functions performed:** Further development and maturation of solution concept, technology maturation (early stage), RDT&E activities, product maturation (late stage). **Adoption functions performed:** Limited fielding, fielding at scale. **Enabling functions performed:** Funding and investment, coordination, navigation and relationships, oversight, hard infrastructure to support development.
How they do their CTP functions	**Technology scouting:** DIU has dedicated engagement teams (one for defense and one for commercial markets) that scan the defense environment for important challenges and the commercial marketplace for potential solutions. DIU uses the commercial solutions opening (CSO) process to accomplish its technology scouting function.[f] The CSO is a competitive and simplified business process to solicit potential solutions to DoD problems and to receive and evaluate proposals. In general, DIU is looking for relatively mature commercial technology or products with the potential to address identified DoD problems. **Network development:** DIU uses its defense and commercial engagement team to maintain relationships with potential DoD customers and commercial technology firms.

Problem identification, validation, and curation: DIU works with DoD components (services, CCMDs, defense agencies) to identify problems and capability needs. It solicits defense problems from across DoD and evaluates whether there is commercial technology with the potential to address the need. As a joint organization, DIU focuses on identifying problems that are shared across multiple DoD components for the broadest impact. DIU translates those capability needs into broad problem statements designed to encourage commercial firms to submit proposals (through its CSO process). Problems are stated as capability needs (what needs to be accomplished) not specific requirements (how to accomplish). Military personnel posted at DIU act as liaisons to facilitate problem identification. The defense engagement teams interact with program offices to better understand the problem being addressed and help determine whether commercial technologies (leveraging commercial RDT&E investment) would be useful.[g]

Solution concept generation: As part of the CSO process, DIU issues an area of interest and invites companies through its website to propose solutions to identified DoD problems. If interested, companies send a short brief about their technology and how it could apply to a particular problem statement. DIU reaches out to companies within 30 days if DIU is interested in scheduling a pitch. Companies that DIU assesses to have made the best pitch are then invited to submit a full proposal. DIU focuses on six technology areas: artificial intelligence, autonomy, cyber, energy, human systems, and space. Each is managed as a portfolio. In most cases, DIU is looking for commercial technology or products that are fairly mature (higher on the TRL scale) and that can be tailored for defense application and potentially fielded within 24 months. The commercial engagement team's access to business intelligence helps determine whether there is an active commercial niche associated with a problem area and plays a role in attracting commercial companies to solve DoD problems.[h]

Matching problems to potential solutions: DIU matches (and then prototypes—see below) commercial technology solutions to address DoD capability needs. "We solicit problem statements based on what we gather by collaboration with the DoD partner. We have commercial executives within DIU that understand the commercial marketplace and what's out there. We have [military personnel] who know what DoD is looking for. If there's a fit and a technology out there that could [solve] our problem, we post the problem set."[i] DIU competitively selects companies with solutions that best fit a stated capability need.

Further development and maturation of solution concepts: To support technology development and further mature the technology and its conceptual application, DIU leverages OTAs and awards prototype agreements. In February 2019, the OUSD/A&S delegated OTA to DIU.[j] In FY 2021, DIU's average time to award a prototype contract was 137 business days.[k] However, the turnaround time is significantly shorter than the traditional DoD contracting process of 18 months or more.

Technology maturation (early stage): Throughout the contract period, DIU provides general support to the firm(s). DIU emphasizes prototype development and testing to demonstrate the technology's ability to address the capability needed. Successful prototypes may transition to other agreements or production contracts from DoD partners or other government entities.

RDT&E activities: DIU focuses on commercial technologies and products that are relatively mature (higher TRL levels). Prototyping activities are designed to demonstrate that the technology or product meets the DoD partner's capability needs.

Product maturation (late stage): Because the goal is to transition or field capability within 24 months, prototypes are closer to product-representative. Prototype activities are designed to help refine product or system design.

Limited fielding and fielding at scale: DIU addresses transition (fielding) issues from the start of a project, including identifying potential transition partners and requirements that a technology or system will need to meet. DIU's goal is to transition systems—either a follow-on contract by a transition partner (i.e., component or PEO) or direct fielding to operation units—within 24 months. Transition is measured by a transition partner taking ownership and management of the project, which includes plans for fielding. DIU is not the primary end customer for the commercial technology solutions (although it has transitioned a number of technologies for its own use).

Funding and investment: DIU uses its CSO process to engage commercial firms and fund prototype development and demonstration efforts. The majority of DIU OTs are structured as firm fixed-price contracts. DIU has an in-house contracting office staffed by agreements officers warranted to issue OTs, which facilitates the agreements process. DIU identifies DoD partner or user funding from those submitting problems.

Coordination: DIU provides general (program management) support to DoD customers and firms through the period of performance with a goal of minimizing the disruption to a company's work while meeting the government's needs, especially meeting milestones, oversight of testing, and contracting.[l]

Navigation and relationships: DIU uses its defense and commercial engagement teams to maintain relationships with DoD partners and commercial technology firms. Specifically, DIU leverages expertise across DoD and the U.S. government (i.e., the Department of Homeland Security and NASA) for technical diligence and has former commercial executives and reservists from industry to help identify the best commercial technology.

Oversight: DIU oversees NSIN and NSIC, though those entities operate independently from DIU.

Hard infrastructure to support development: DIU can leverage DoD resources like test ranges and certification authorities to support prototype demonstration.

CTP-related outputs	DIU outputs include validated DoD problems, identification of promising commercial technologies, solution concepts generated, technologies matured, and technologies fielded for use by the military. All numbers referenced below are from the DIU, *Annual Report FY 2021.*[m] • DIU's goal is to move from problem identification to a complete prototype contract award in 60 to 90 days. As of FY 2021, 41% of projects yield at least one successfully transitioned commercial solution. • In 2021, DIU increased by 10% the average number of proposals received per solicitation (43) compared with FY 2020. • In 2021, DIU, in collaboration with DoD partners seeking commercial solutions to national security challenges, initiated 26 new projects (a 4% increase), bringing its total project count up to 119 (including active and completed). As a result, DIU awarded 56 prototype OT agreements to companies. Most of the OT agreements are with small businesses or nontraditional defense contractors. Since DIU began prototype activities in 2016, it has awarded 279 prototype OT agreements to companies across 28 U.S. states and eight foreign countries. Eight of those transitioned in FY 2021.
Definition of success	Key metrics of success for DIU include whether technology transitions to an end user, whether solutions scale across DoD or other government organizations, speed to contracts, and overall impact, which "becomes harder to measure."[n] According to DIU's definition, "a commercial solution transitions when the prototype successfully completes and results in a production or service contract with a DoD or U.S. Government entity. A transition enables DoD to field a product or solution in an operational environment supporting warfighters."[o]

DIU is also focused on growing the national security innovation base. Between June 2016 and September 2021, 33% of awardees were first-time vendors, 86% met the criteria for nontraditional firms, and 73% were small businesses.[p]

DIU published a total of 26 new solicitations to its website in FY 2021 alone and received an average of 43 proposals per solicitation. The average number of business days to award a prototype contract in FY 2021 was 137. Finally, the investment DoD made in DIU six years ago is bearing fruit—evidenced by its 47 transitions (as of June 2022), representing $3.9 billion in contract ceiling (~$84 million average contract ceiling per transition) since 2015. These 47 transitions led to billions of dollars in additional DoD revenue opportunities from non-DIU contracts.[q]

CTP-related outcomes	The following are examples of success stories—that is, successfully transitioned projects:[r] • Generative Modeling of Hypersonic Missile Trajectories: Missile Defense Agency (MDA) requires accurate and timely models of incoming missile trajectories in order to simulate real-world performance of nonballistic and hypersonic missiles. Physics-based trajectory models take many months to develop and generate limited numbers of trajectories. In May 2021, the MDA awarded C3.ai an OT production contract after the company successfully completed a prototype with DIU to simulate the real-world trajectories of nonballistic and hypersonic missiles. MDA will leverage C3.ai's AI tool, which provides a multifaceted development studio for data integration, operations, and security, to expand simulation capabilities for nonballistic and hypersonic missiles. • Installation Counter-UAS: Anduril received a five-year OT production award in June 2021 for a system that "combines the latest in artificial intelligence techniques with sensor technology to enhance physical security through automated detection, identification, and defeat of objects of interest while reducing manpower requirements." Transition partners include U.S. Southern Command, U.S. Marine Corps, Defense Threat Reduction Agency, Naval Air Systems Command, Customs and Border Protection, and U.S. Central Command. • Cyber Asset Inventory Management: IntelliPeak Solutions received an OTA production award in September 2021 for its Axonius platform. The prototype proved it can fill a gap in DoD's existing tools by integrating "data from across the DoD enterprise to understand all deployed assets, their current software and firmware, and their configurations." Transition partners included the Defense Information Security Agency and the Joint Service Provider.

[a] DIU, "About," webpage, undated a.

[b] DIU information to authors, July 11, 2022.

[c] DIU, *Annual Report FY 2021*, p. 4.

[d] DIU, undated a; National Security Innovation Capital, "About," website, undated.

[e] DIU, *Annual Report FY 2021*.

[f] DIU (experimental), "Commercial Solutions Opening," white paper, undated b; Shay Assad, director, OUSD/A&S Defense Pricing, "Class Deviation—Defense Commercial Solutions Opening Pilot Program," memorandum, June 26, 2018.

[g] Interview 40, Interviewee 52, November 3, 2021.

[h] Interview 40, Interviewee 52, November 3, 2021.

[i] Interview 40, Interviewee 52, November 3, 2021.

[j] DIU, *Annual Report 2019*, p. 5.

[k] Interview 11, Interviewee 12, August 25, 2021.

[l] Interview 40, Interviewee 51, November 3, 2021.

[m] Starting in 2021, DIU changed its annual reporting period from calendar year to fiscal year that begins October 1.

[n] Interview 11, Interviewee 12, August 25, 2021.

[o] DIU, *Annual Report FY 2021*, p. 8.

[p] DIU, *Annual Report FY 2021*; Interview 40, Interviewees 12 and 51, November 3, 2021.

[q] DIU, *Annual Report FY 2021*; DIU information to authors, July 8, 2022.

[r] All the examples listed are taken from DIU, *Annual Report FY 2021*, p. 11.

Doolittle Institute

Snapshot of organization	**Founded:** Established in 2012 as a 501(c)(3) nonprofit.[a] **Mission and focus:** The Doolittle Institute (DI) supports the Air Force Research Laboratory's Munitions Directorate (AFRL/RW) by "working to license and commercialize AFRL/RW technologies in the private sector, enable rapid technology delivery to the warfighter, identify and foster new R&D partnerships, and develop AFRL's current and future workforce."[b] **Organizational statistics:** DI is part of a network of five AFRL innovation institutes (the other four are AFRL New Mexico, the Griffiss Institute, BRICC, and the Wright Brothers Institute).[c] DI has 15 staff members. **How it operates:** DI primarily engages the CTP through two services: • Technology Transfer (T2): The Air Force T2 program seeks to increase the economic competitiveness of industry. DI facilitates the Air Force T2 program for AFRL/RW through CRADAs, commercial test agreements, information transfer agreements, material transfer agreements, and patent license agreements.[d] DI also manages Innovation Discovery Events (IDEs) for AFRL/RW scientists and engineers. IDEs facilitate labs identifying technologies to protect through patents and disclose to the public for potential commercialization and technology transfer under license. IDEs also help AFRL/RW researchers identify new inventions and promising commercial applications for those inventions in nontraditional industries. • Innovation and Collaboration: DI hosts and supports regional entrepreneurship and innovation efforts, SBIR Days, SBIR Sprints, and presentations and workshops. **Primary customer:** AFRL/RW.
Link to CTP	**Core functions performed:** Technology scouting, customer discovery, network development, technology maturation (early stage). **Enabling functions performed:** Coordination, navigation and relationships, information sharing and reporting, provision of hard infrastructure, provision of soft infrastructure.
How they do their CTP functions	**Technology scouting:** Unlike most DoD organizations in the innovation ecosystem, DI's primary mission is to transfer research and technology from AFRL to industry for commercialization, rather than find commercial companies or solutions for DoD problems.[e] That said, DI wants to shift to also focus on finding commercial technologies to address Air Force problems.[f] DI holds university days, which are events where researchers can submit white papers for AFRL scientists and engineers to review and potentially receive seedling funding to support the researchers' further development of their research. **Customer discovery:** AFRL/RW tasks DI with identifying nontraditional companies, particularly small businesses, and incentivizing them to participate in the DoD ecosystem. **Network development:** DI hosts industry days, collider events, and SBIR Days to partner with AFRL and connect small businesses with Air Force opportunities in the SBIR and STTR programs. DI helps small businesses with presenting their research and products to AFRL. **Technology maturation (early stage):** DI supports early-stage technology maturation by connecting small businesses with SBIR phase I contract opportunities and introducing them to customers that can sponsor a SBIR phase II contract.

DI hosts Grand Challenges, which are competitions based on a selected topic area (e.g., machine learning), with a $1 million prize for the winner. A Metal Additive Manufacturing Grand Challenge consisted of three phases. In the first phase, companies were required to submit a four- to six-page paper. The second phase required the company to develop and demonstrate its approach to manufacturing a coupon (a small piece of metal) to specifications and requirements set by AFRL. The third phase required the company to demonstrate its approach by manufacturing a steel cylinder with complex requirements (similar to a munition) specified by AFRL.[g] Companies are required to transfer government purpose rights to the government to use the technology as a condition of winning a prize, but companies retain IP rights.[h]

Coordination: DI establishes CRADAs between AFRL and companies or research institutions.[i]

Navigation and relationships: DI holds IDEs to help AFRL/RW scientists and engineers with the patent process. DI provides invention discovery, claims improvement to get to a commercially viable patent application, and commercialization exploration services depending on where scientists and engineers are in the patent process. Before patent disclosure, DI helps them "clearly and concisely articulate what is new and different about their innovation." Then DI helps the performers improve invention disclosures before applying for a patent so that their chances of obtaining a broader, commercially viable patent increases. After applying for a patent, DI helps scientists and engineers communicate the commercial use cases for their innovations with the goal of improving marketing to industry.

Information sharing and reporting: DI provides a link to the AFRL/RW intellectual property portfolio on TechLink.[j] This portfolio provides the public with a list of technologies available for licensing and commercialization with a description of the technology and a link to the patent.

Provision of hard infrastructure:

- DI provides hard infrastructure such as laboratory and test equipment through commercial test agreements (CTAs) under which "a defense laboratory may make available to any person or entity, at an appropriate fee, the services of any government laboratory for the testing of materials, equipment, models, computer software, avionics systems, antennas and other items."[k]
- DI uses information transfer agreements (ITAs) to provide Air Force–developed software to other partners to commercialize or help meet Air Force problems. An ITA allows DI to "share internally developed [Air Force] software related to design or manufacturing activities with other entities. Software executable files and/or source code may be shared under the agreement with industry and academic partners."[l]
- DI establishes material transfer agreements (MTAs) to "allow the government to share material and equipment with other organizations for the purpose of cooperative research and development."[m]

Provision of soft infrastructure: DI supports commercialization of Air Force technology through patent license agreements (PLAs). A PLA is "a contract between a licensor (e.g., the holder of a patent such as the USAF) and a licensee (e.g., an industry partner) which ensures the licensee that the licensor will not sue the licensee for patent infringement."[n]

CTP-related outputs	DI uses CTAs, CRADAs, ITAs, MTAs, and PLAs to allow for-profit and nonprofit companies to gain access to AFRL resources, perform cooperative research with AFRL, use AFRL equipment for cooperative research, and hire AFRL to perform specialized testing.[o]
Definition of success	Doolittle Institute provides capabilities to achieve operational objectives in support of the National Defense Strategy. It aims to "support AFRL to collaboratively work with other DoD entities, industry, and academia to accelerate the ideation, conceptualization, and prototyping of digital, low-cost, and high-tech capabilities for the warfighter." Successes include the discovery of new innovations, ideation of problems and solutions, and ultimately the integration of new capabilities as components of systems, particularly to meet the primary sponsor's needs in advanced munitions.

CTP-related outcomes	Commercialization of AFRL/RW S&T activities by the private sector after technology transfer is achieved by • finding new industries that DI can transfer technology to from AFRL/RW[p] • providing funding to small businesses through grants • connecting small businesses to AFRL opportunities, primarily through SBIRs and STTRs.

[a] Doolittle Institute was later rebranded as DEFENSEWERX in 2017. The Air Force Research Laboratory Munitions Directorate (AFRL/RW) retained the name "Doolittle Institute" for the Northwest Florida DEFENSEWERX facility located in Niceville, which operates under a partnership intermediary agreement (PIA) between DEFENSEWERX and AFRL/RW. See Doolittle Institute, "AFRL Innovation Institutes," webpage, 2021a.

[b] Doolittle Institute, 2021a.

[c] Doolittle Institute, "Our History," website, undated b.

[d] Doolittle Institute, "Technology Transfer," webpage, 2021b.

[e] Doolittle Institute, 2021b.

[f] Interview 29, Interviewee 39, September 29, 2021.

[g] Doolittle Institute, "Grand Challenge: Metal Additive Manufacturing," webpage, undated a.

[h] Government purpose rights allow the government to share technical data internally without restrictions and may disclose such data to private entities for noncommercial purposes, such as a contractor requiring access to perform government-contracted services. See "Rights in Technical Data—Noncommercial Items, 252.227-7013," *Defense Federal Acquisition Regulation Supplement*, February 2014.

[i] Doolittle Institute, 2021b.

[j] TechLink, "Air Force Research Laboratory Munitions Directorate," webpage, undated.

[k] Doolittle Institute, 2021b.

[l] Doolittle Institute, 2021b.

[m] Doolittle Institute, 2021b.

[n] Doolittle Institute, 2021b.

[o] Doolittle Institute, 2021b.

[p] Interview 29, Interviewee 39, September 29, 2021.

Joint Rapid Acquisition Cell

Snapshot of organization	**Founded:** Established in September 2004 by Paul Wolfowitz, then–deputy secretary of defense, in order to "break down the institutional barriers that prevent timely and effective joint warfighting support."[a] **Mission and focus:** JRAC "coordinates urgent and emergent acquisitions to meet the needs of combatant commanders and interagency partners by refining requirements and harnessing authorities best suited for rapid acquisition. Within DoD, JRAC provides Warfighter Senior Integration Group oversight for ongoing counterterrorism operations and coordinates Combatant Commander-wide joint urgent and emergent operation needs.[b] **Organizational statistics:** Per statute, DoD may use Rapid Acquisition Authority (RAA) to transfer or reprogram up to $800 million per fiscal year to meet urgent needs. **How it operates:** JRAC reports to the Assistant Secretary of Defense for Acquisition and is directly responsive to CCMD operational needs. When an urgent or emergent need arises, the sponsor command recommends the use of a Joint Urgent Operational Need (JUON) or Joint Emergent Operational Need (JEON) authority. If the need cannot be met by an existing service or program of record, the JUON/JEON is submitted to the combatant commander (CCDR) for certification and prioritization. If the CCDR agrees to certify the JUON/JEON, it is then submitted to the Joint Staff and JRAC. Pursuant to the Joint Staff's recommendation, JRAC "designates or declines the JUON as [an immediate warfighter need (IWN)] within 14 days of submission to JRAC."[c] JRAC is responsible for

identifying a funding source and the appropriate service lead for transition. Ultimately, JRAC transitions the JUON/JEON to an appropriate DoD component for action but continues to monitor until the IWN is fulfilled. JRAC's objective is to designate an urgent operational need as an "IWN" within 48 hours of receipt and to resolve the IWN on a timeline of between 120 days and two years.[d]

Primary customers: CCMDs.

Link to CTP	**Identification functions performed:** JRAC provides problem identification, validation, and curation. **Development functions performed:** JRAC provides further development and maturation of solution concept. **Adoption functions performed:** Limited fielding. **Enabling functions performed:** JRAC provides policies, guidance; funding and investment; coordination; oversight.
How they do their CTP functions	**Problem identification, validation, and curation:** As noted under *How it operates*, JRAC receives a JUON/JEON request directly from the CCDR, subject to validation by the Joint Staff. JRAC acts as the "single point of contact" within OSD for JUONs.[e] Thus, JRAC does not perform problem identification, which is done by the CCDRs through JUONs/JEONs. However, it is part of the validation process, along with the Joint Staff, for JUONs. JRAC, together with the given functional capabilities board (FCB) within the Joint Staff, works with the originating command to refine its JUON request. It is important to note that the services have their own unique set of processes distinct from JRAC's that are used for service-specific urgent operational needs. **Further development and maturation of solution concept:** JRAC does not directly identify companies or solutions. However, it does work with the originating command to refine the material solution and coordinate with the lead service to develop, acquire, and field the solution. After the Joint Staff validates the CCDR-initiated urgent operational need, JRAC works with the appropriate FCB within the Joint Staff and the originating command to refine the material solution.[f] **Adoption:** JRAC facilitates acquisition of systems to address urgent needs through its rapid acquisition authorities under DoDI 5000.78. In particular, JRAC coordinates the approval process for JUONs, identifies funding, and designates the lead service to be responsible for rapidly developing, acquiring, and fielding the material solution. **Policy and guidance:** Per DoDM 5000.78, the executive director of JRAC is delegated the responsibility for making all RAA recommendations and documenting requirements, sponsors, and resources.[g] Additionally, JRAC is responsible for preparing supporting documentation for any RAA budget requests to Congress. **Funding and investment:** JRAC is responsible for identifying the funding source for an urgent operational need. It can accomplish this through the Rapid Acquisition Fund, as well as reprogramming requests. Per DoDM 5000.78, JRAC is responsible for preparing all RAA budget documentation. However, interviewees clarified that JRAC is not involved in the contracting process.[h] **Coordination:** Although it is unclear to what extent JRAC interacts with other DIOs, the organization maintains close contact with the CCDRs and the Joint Staff. It also facilitates interactions between initiating commands and the service to which a solution will ultimately be assigned. **Oversight:** Once a given urgent operational need has been transferred to the designated service, JRAC tracks and monitors the progress, acting as a point of contact between the service and CCDR.
CTP-related outputs	The immediate outputs are a JUON or JEON to be fulfilled by a particular DoD component or service; funding transfers through the RAA; and rapid acquisition policies and guidance are provided by the JRAC advisory board.

Definition of success	JRAC seeks to rapidly field material solutions to meet urgent needs within 120 days to two years. JRAC therefore defines success in terms of both the speed at which it was able to facilitate fielding and whether the solution met the originating command's operational need. Since JRAC's rapid acquisition funding is ad hoc, typically requiring reprogramming from other program elements, it also views transition to a program of record or adoption within a service's acquisition portfolio as a success in the context of providing the CCDR customer with funding stability.
CTP-related outcomes	Interviewees pointed to two key success stories: a system for transporting personnel infected with COVID-19, and counter-unmanned aircraft systems (c-UAS).[i] Regarding the safe transport system, one interviewee stated the following: "The Air Force moves infected personnel in container beds with no room for troops that get hurt in combat. We developed a TIS [Transportation Isolation System] and then went to the containerized pressurized systems to move COVID-infected personnel. The technology was there [originally] to protect people from CBRN [chemical, biological, radiological, or nuclear hazards]. It was not that far removed from them for how to take care of people in the pandemic."[j] Regarding c-UAS, the interviewee stated: "The best example is with c-UAS. There were multiple solutions. . . . The Army created an executive agent to take great ideas . . . best of breed. There were six c-UAS iterations. We'll continue those. That's an enduring need, which will be with us for many moons to come."[k]

[a] Robert L. Buhrkuhl, "When the Warfighter Needs It Now," *Defense AT&L*, November–December 2006, p. 29.

[b] In particular, JRAC "acquires goods and services as rapidly and effectively as possible for the Federal Government in support of the American public" through the Defense Assisted Acquisition (DA2) program. JRAC provides DA2 support to the interagency "when the magnitude of the Government's requirements overwhelms the lead response agency(ies)." See Office of the Assistant Secretary of Defense, Acquisition, "Joint Rapid Acquisition Cell," webpage, undated.

[c] DAU "Joint Rapid Acquisition Cell," webpage, undated a.

[d] Joint Rapid Acquisition Cell, "Meeting Warfighter Needs for the Asymmetric Threat," presentation, NDIA Gun and Missile Systems Conference, April 25, 2007.

[e] Buhrkuhl, 2006.

[f] Michael W. Middleton, *Assessing the Value of the Joint Rapid Acquisition Cell*, Thesis, Naval Postgraduate School, December 2006.

[g] DoD Manual 5000.78, *Rapid Acquisition Authority (RAA)*, Washington D.C.: U.S. Department of Defense, March 20, 2019.

[h] Interview 15, Interviewee 17, September 3, 2021.

[i] Interview 15, Interviewee 17, September 3, 2021.

[j] Interview 15, Interviewee 17, September 3, 2021.

[k] Interview 15, Interviewee 17, September 3, 2021.

National Security Innovation Network

Snapshot of organization	**Founded:** NSIN was founded in October 2016 as the MD5 National Security Technology Accelerator.[a] In 2019, MD5 was renamed the National Security Innovation Network.[b] **Mission and focus:** NSIN encourages innovators from academia and the private sector to solve DoD problems and supports the identification and development of innovative tools and solutions that address DoD problems.[c] **Organizational statistics:** As of FY 2022, NSIN has a staff of 62 personnel, including government employees and contractors. Outside the National Capital Region, NSIN has a regional network of 11 regional directors supported by 16 university program directors at various universities across the country.[d]

NSIN funding was $5 million in FY 2016, $25 million in FY 2017, $25.5 million in FY 2018, and $15 million in FY 2019. It received $40 million as a congressional interest program in FY 2020 and in FY 2021.[e] FY 2022 is the first year that NSIN was funded as a program element through the FYDP; the president's FY 2022 budget request to support NSIN was $21.27 million.[f]

Section 219 of the FY 2021 National Defense Authorization Act (NDAA) directed the Secretary of Defense to establish "an activity" to conduct many of the activities that NSIN undertakes.

Reporting structure: NSIN is a program office within OUSD/R&E that reports through DIU to the Under Secretary of Defense for Research and Engineering.

How it operates: NSIN has three activity portfolios (National Service, Collaboration, and Acceleration) that together manage 16 programs. We are focused on the *Acceleration* portfolio of programs because of the nexus to the CTP.[g]

Primary customers: NSIN works across the DoD and services, identifying and "curating" problems from potential DoD mission partners/end users at any level within DoD (e.g., a single unit, base, division, or combatant command), and it engages potential partners from diverse communities (including academic and venture communities).

Link to CTP	**Identification functions performed:** Technology scouting; customer discovery; problem identification, validation, and curation; solution concept generation; network development; matching problems to potential solutions **Development functions performed:** Further development and maturation of solution concept, technology maturation (early stage). **Enabling functions performed:** Funding and investment, navigation and relationships.
How they do their CTP functions	NSIN's *Acceleration* portfolio is the most closely linked to the CTP.[h] Through Acceleration programs, NSIN seeks to develop and "improve the viability and sustainability of dual-use ventures" that "address [DoD] capability gaps" and "accelerate the adoption of novel concepts and solutions."[i] The portfolio includes the following programs: • *Foundry*, formerly known as DIA (Defense Innovation Accelerator): NSIN matches DoD lab and FFRDC technologies to entrepreneurs who seek to commercialize the technologies. Through the program, entrepreneurs are matched with technologies and provided with instruction on how to form and run a business.[j] They are also provided with mentorship from private investors and DoD officials who can help them learn how to interact with the government. The entrepreneurs assess the technologies, their potential customers, markets, competitors, and commercialization opportunities; create new companies (start-ups) to commercialize DoD lab technologies; and negotiate with defense labs for licensing agreements.[k] **CTP functions performed through *Foundry*:** customer discovery. • *Propel*: NSIN provides early-stage companies with promising TRL 4–6+ technologies with training, support to identify potential revenue sources, and follow-on contracts to support their maturation and development. Early-stage companies with solution concepts that map to identified DoD problems participate in a customized commercial accelerator and receive mentorship; learn how to engage with DoD; and explore use cases, customers, potential revenue sources, investors, and non-dilutive funding mechanisms.[l] The most promising early-stage dual-use ventures may receive a follow-on contract to continue to mature and develop their small business and solution concepts. **CTP functions performed through *Propel*:** technology scouting; customer discovery; problem identification, validation, and curation; solution concept generation; network development; matching problems and solutions; funding and investment; navigation and relationships.

- *Vector*: Vector helps promising NSIN-alumni teams form new dual-use ventures to develop and sell their technologies to DoD and the commercial market. Selected teams participate in a ten-week accelerated learning program to learn entrepreneurship fundamentals, how to do business with DoD, and how to navigate different markets. The winning team receives a $25,000 contract that can be used to fund better refinement and maturation of the product-market fit. The contract is not seen as the main motivator for participation in this programming; instead, the opportunity to network and gain access to DoD officials is the "main pull."[m]
  CTP functions performed through *Vector*: funding and investment; navigation and relationships.
- *Starts*: NSIN convenes a "pitch" day to introduce DoD customers and end users to new technology solutions. NSIN partners with a DoD customer to develop a problem statement that describes an unmet need. Early-stage companies with solution concepts that map to those problems participate in a one-day pitch event where they present their solution concepts to DoD customers, venture capital/private investors, and academics. The Starts program works with the DoD problem owner to identify the available, and optimal, incentive structure for the most promising dual-use, early-stage ventures. This can take the form of receipt of a financial prize awarded through a prize authority, access to a non-dilutive contracting pathway for the transition and development of solutions, or guaranteed declarations of interest for the SBIR process.[n]
  CTP functions performed through *Starts*: technology scouting; customer discovery; problem identification, validation, and curation; solution concept generation; network development; matching problems and solutions; funding and investment; navigation and relationships.
- *Emerge Accelerator (pilot)*, a successor to NSA2 (pilot) (National Security Academic Accelerator): NSIN seeks to establish technology transfer and transition partnership with universities, leveraging their existing intellectual property, to create new dual-use ventures that can meet DoD's urgent and critical areas of technology development. The Emerge Accelerator provides funding to support university students, faculty, and other staff and/or companies (ideally at the prototype stage) to develop technology and form ventures for commercial and DoD applications.[o]
  CTP functions performed through *Emerge*: technology scouting; funding and investment

Technology scouting: NSIN searches for technologies and early-stage ventures through targeted web-based searches of relevant government and private-sector databases and sources. Further, NSIN personnel located in targeted cities and universities around the United States (in NSIN's regional network team) conduct outreach and identify early-stage, potentially dual-use ventures and technology development trends in and near targeted cities and universities.[p]

Customer discovery: Across its *Acceleration* portfolio programs, NSIN convenes and connects innovators and prospective DoD customers so they can explore DoD use cases.

Problem identification, validation, and curation: NSIN conducts outreach to and engages potential sponsors, partners, and end users from DoD headquarters, major commands, and at the local base level through its regional network team (RNT). Through the RNT, NSIN seeks to identify "core problems down to the end-user level and bring them into [NSIN] programming."[q] NSIN "curates" problems, seeking to identify problems and end users that are a good fit for specific NSIN programs— namely, end users with budgets and authority to adopt a solution from a commercial organization and a willingness to support programming efforts and to help develop a transition plan.[r]

Solution concept generation: NSIN supports the development of technology ideas that map to identified DoD problems by providing early-stage ventures with opportunities to participate in customized accelerators, to receive mentoring on how to engage with DoD customers, and to compete for streamlined contracting pathways that allow for technology development. (See the Starts and Propel program summaries above.)

Technology development (early stage): Through its Maker program (part of the Collaboration portfolio), NSIN provides funding and/or technical expertise to construct prototypes of novel solutions that were developed and/or identified through other NSIN programming.[s]

Network development: NSIN seeks to foster networks of innovators and DoD end users. Internally, NSIN's RNT provides the organizational and management structure for the execution of most NSIN programs. The RNT structure is built around hub cities chosen based on relative proximity to a concentration of DoD customers and tier 1 research universities,[t] as well as the relative strength or assessment of the commercial innovation/venture ecosystem. These hubs are supplemented by spokes anchored by university partners that help engage academic problem solvers. Using these physical hubs and spokes, NSIN engages entrepreneurs, dual-use ventures, major commands, and units and bases throughout the country to enable problem identification and solution matching.[u] NSIN is seeking to create and populate a digital Defense Innovation Network (DIN) platform through which innovators and DoD customers can connect and collaboratively solve problems.

Matching problems and solutions: NSIN identifies and curates DoD customers and problems, innovators, and early-stage ventures and matches them through its Acceleration portfolio programming (Foundry, Propel, Starts).

Further development and maturation of solution concept: NSIN programming provides guidance, training, and mentorship (Vector, Propel, and Foundry programs). It also assesses commercialization potential (Maker program in the Collaboration portfolio) and funding to support development and maturation of solution concepts (Maker program). Innovators and early-stage ventures explore how to assess the commercialization potential of the inventions, licensing, and how to pursue SBIR/STTR, venture capital, and other funding opportunities.[v]

Funding and investment: NSIN provides limited funding to support development and maturation of solution concepts. NSIN also helps early-stage ventures secure DoD and commercial funding by connecting early-stage ventures to prospective DoD customers, VCs, private investors, and potential DoD end users who may be able to provide contracting opportunities. In addition, NSIN identifies additional sources of funding (e.g., SBIR/STTR) and works with VCs, commercial accelerators, and DoD customers who are able to provide staff and to teach and mentor ESVs and entrepreneurs on how to engage with and secure funding from government and commercial investors.

Navigation and relationships: NSIN provides training, mentorship, and introductions to help early-stage ventures and innovators understand and navigate DoD and build relationships with DoD end users and prospective customers.

CTP-related outputs	Key outputs include the following: • DoD problems are identified and curated. • Early-stage ventures are identified, created, and matured. • Solution concepts are identified, created, and matured. • Customers/problems are matched to EVSs and solution concepts. • Funding sources and "transition pathways" are identified for ventures and concepts. • Innovators and entrepreneurs are motivated to solve DoD problems; they are educated and allowed to discover and partner with DoD customers and form dual-use ventures.

Definition of success	NSIN seeks to meet five enterprisewide objectives and tracks a number of "key results" for each:[w]
	
	1. Generate solutions to national security problems through collaborative interaction with DoD, academia, and the venture community. Related to this objective, NSIN tracks results including
	  • number of ventures and individuals that participate in NSIN programming and activities
	  • DoD mission partner satisfaction rating.

2. Expand the National Security Industrial Base (NSIB). Related to this objective, NSIN tracks results for
   - number of first-time individual entrants into NSIN programming.
3. Mature the NSIN network. For this objective, NSIN tracks results including
   - percentage of NSIN alumni who continue to engage in NSIN programming and activities
   - number of commercial ventures that transition into a dual-use market through NSIN programming or organic problem solving/outreach.
4. Facilitate transfer and transition of U.S. government and university inventions. For this objective, NSIN tracks the
   - number of government or university inventions that are licensed by NSIN network members as a result of NSIN programming.
5. Increase the speed and regularity of transition of solutions identified or generated through NSIN programming and activities. In connection with this objective, NSIN tracks results including
   - percentage of solutions generated through NSIN programming that receive follow-on funding from another U.S. government organization
   - percentage of solutions generated by NSIN programs that are in operational use or implemented by DoD.

CTP-related outcomes

NSIN tracks the following CTP-related outcomes:
- number of commercial ventures that transition to dual-use markets through NSIN programming
- percentage of solutions generated by NSIN programs that are in operational use by DoD.[x]

---

[a] U.S. Department of Defense, "MD5—A New Department of Defense National Security Technology Accelerator—Officially Launches with Disaster Relief Hackathon in New York City," press release, October 14, 2016.

[b] NSIN, "MD5 Adopts New Name to Reflect Refined Mission," webpage, May 6, 2019.

[c] "NSIN Overview Brief," on file with authors, undated; NSIN, "Mission," webpage, 2021; NSIN All Hands Memorandum 2 (Vision and Mission), on file with authors, July 8, 2019. Various "all-hands memos," which are internal NSIN documents, were provided to the authors and used to compile this profile.

[d] NSIN, "Team," webpage, 2021b.

[e] We note that NSIN's FY 2020 request was $25 million, but it received $40 million.

[f] DoD, *Fiscal Year 2022 Budget Estimates: Defense-Wide Justification Book Volume 3 of 5*, May 2021, esp. PE 0603950D8Z: National Security Innovation Network, p. 371.

[g] In addition to Acceleration, NSIN runs two other portfolios of programs: the *National Service* portfolio and the *Collaboration* portfolio. National Service programs "create new opportunities for national security service for those who might not otherwise participate in national security innovation" (NSIN, "Mission," webpage, 2021a). Programs in this portfolio pair student teams with DoD to solve problems as part of independent study, identify mid-career academic experts with desirable backgrounds, facilitate access to certain hiring authorities for hard-to-reach talent, operate fellowships, or connect early-career STEM professionals with DoD units. The Collaboration portfolio is meant to "increase the intellectual diversity of the DoD by bringing together innovators from defense, academia, and the venture community to solve national security problems" ("NSIN Overview Brief," on file with authors, undated). Collaboration programs include Hacks, Hacking for Defense (H4D), and Bootcamp, which bring together civilian and military innovators to solve DoD problems or, as part of a university course, encourage students to solve problems and think entrepreneurially. Programs also offer funding or technical expertise to construct prototypes, give service members the opportunity to provide direct feedback to senior leaders through virtual innovation challenges, and teach service members skills to solve problems that directly affect their command ("NSIN Overview Brief," on file with authors, undated).

[h] "NSIN Overview Brief," on file with authors, undated.

[i] NSIN, "Accelerator Portfolio RNT [Regional Network Team] Guide with Branding," on file with authors, February 15, 2021.

[j] Interview 19, Interviewee 21, September. 15, 2021.

[k] Interview 19, Interviewee 21, September 15, 2021; NSIN, "Foundry," webpage, undated b.

[l] NSIN, "Accelerator Portfolio RNT Guide with Branding," on file with authors, February 15, 2021.

[m] Interview 6, Interviewee 7, August 11, 2021; NSIN, "Accelerator Portfolio RNT Guide with Branding," on file with authors, February 15, 2021.

[n] NSIN, "Accelerator Portfolio RNT Guide with Branding," on file with authors, February 15, 2021; Mission Model Canvas, Starts, designed by Joey Clark (Starts MMC v4_0), on file with authors, April 25, 2020; other information provided by NSIN to authors, June 30, 2022.

[o] NSIN, "Emerge Accelerator," webpage, undated a.

[p] NSIN All Hands Memorandum 3, on file with authors, September 3, 2019; Interview 1, Interviewee 1, Jul. 20, 2021.

[q] Interview 1, Interviewee 3, July 20, 2021.

[r] Acceleration Overview, PowerPoint presentation, on file with authors, July 15, 2021.

[s] "NSIN Overview Brief," on file with authors, undated; other information provided by NSIN to authors, July 12, 2022.

[t] Tier 1 refers to "R1 Doctoral Universities—Very High Research Activity" according to the Carnegie Classification of Institutions of Higher Education, as developed by the Carnegie Foundation for the Advancement of Teaching and the American Council on Education (ACE).

[u] NSIN All Hands Memorandum 3, on file with authors, September 3, 2019.

[v] Acceleration Overview, PowerPoint presentation, on file with authors, July 15, 2021.

[w] The objectives and key results listed here are derived from NSIN, "Change 2 to NSIN All Hands Memorandum 4," on file with authors, October 1, 2021.

[x] NSIN, "Change 2 to NSIN All Hands Memorandum 4," on file with authors, October 1, 2021.

NavalX

Snapshot of organization	**Founded:** Established in Alexandria, Virginia, in February 2019. **Mission and focus:** NavalX's mission is to function as the bridge between Navy organizations with defense challenges and innovators with solutions in industry, academia, and the Department of the Navy (DoN).[a] NavalX is also focused on developing an innovation culture in the Navy through training and education. **Organizational statistics:** NavalX has a core staff of 20, plus six directors at its Centers for Adaptive Warfighting and 17 regional Tech Bridge directors that operate the 15 Tech Bridges across the United States and two overseas locations: one in London and one in Yokosuka, Japan; Tech Bridge projects totaled more than $50 million in 2020.[b] **How it operates:** NavalX operates "Tech Bridges," which are regional platforms intended to create a network of people, ideas, and best practices and helps DoN stakeholders deliver capabilities using an eight-step innovation pipeline (see below for *How they do their CTP functions*).[c] NavalX also runs six Centers for Adaptive Warfighting that teach and employ agile business practices to address military problems, such as using design thinking and scrum techniques.[d] **Primary customers:** Department of the Navy (U.S. Navy and U.S. Marine Corps components).
Link to CTP	**Core functions performed:** Technology scouting; problem identification, validation, and curation; solution concept generation; network development; technology maturation (early stage); RDT&E activities; product maturation (late stage); and transition to fielding. **Enabling functions performed:** Funding and investment, coordination, navigation and relationships.
How they do their CTP functions	NavalX facilitates interactions between Navy commands or units with problems and organizations that can potentially solve them, creating networks and building teams. NavalX specializes in "human-centered problem solving,"[e] which includes training DoN personnel in agile methodologies and design thinking. It delivers capabilities using an eight-step innovation pipeline: scan, source, curate, incubate, prototype, validate, transition, and sustain.

Technology scouting: NavalX performs market surveys, technology scanning, and government lab scanning to learn how others address similar problems.[f]

Problem identification, validation, and curation: Tech Bridges search for local and regional DoN problems; DoN personnel can also come to Tech Bridges with problems.[g] Each Tech Bridge represents the installations within its geographic region. NavalX curates problems by defining stakeholders, problem owners, and requirement holders; identifying funding; and refining the problem statement.[h]

Solution concept generation: NavalX uses non-FAR-based contracting and agreement mechanisms such as prize challenges and technology transfer authorities (e.g., PIAs) to attract a wide pool of expertise and solution sets to support the development of technology, ideas, and concepts for DoN problems.[i] Partners include academia, industry, nonprofit organizations, and government agencies.

Network development: NavalX uses Tech Bridges to develop regional innovation networks that connect DoN personnel with problems to companies.[j] These networks increase collaboration, knowledge sharing, and innovation between the DoN, Naval Labs, industry, academia, and other military services. NavalX also offers Tech Bridges access to state and local governments[k] within the United States and the Royal Navy in the United Kingdom and Japanese stakeholders in Japan.[l]

NavalX builds informal relationships with nearby commercial companies, research institutions, and state and local governments that can later become formalized through memorandums of agreement, consortia, contracts, and other mechanisms.[m]

Technology maturation (early stage): NavalX creates and iterates minimum viable products (MVPs) with solution developers by employing scrum techniques, gathering input from potential end users through each "spring" to develop features in the solution.[n]

RDT&E activities: NavalX creates and tests prototypes with end users.[o] It works with the DoN personnel/end users to validate solutions on test platforms.[p]

Product maturation (late stage): NavalX provides validated prototypes to acquisition organizations for further maturation and production so that the technologies can transition and solve DoN capability gaps.[q]

Transition to fielding: This is the step in the innovation pipeline that scales the solution. NavalX starts working the handoff between its customer and program offices or users as soon as NavalX begins problem solving with a customer. NavalX finds a market for the solution, promotes and launches that solution, and then trains users and receives feedback and usage data. NavalX creates/works with life cycles for products, predictive maintenance schedules, and obsolescence."[r]

Funding and investment: NavalX leverages prize challenges, OTA consortia, and regional partnerships.[s] However, it does not have contracting authority.[t]

Coordination and navigation and relationships:

NavalX mentors and educates small businesses on how to do business with the Navy, including business and registration requirements to receive government contracts; it also connects small businesses with other departments. For example, a small business may have an idea or technology that does not align with DoN but may be useful to the Army, so NavalX makes that connection.[u]

CTP-related outputs	The main outputs are prototypes and government off-the-shelf solutions that the DoN can alter and leverage as needed.[v]
Definition of success	Success is when the Office of Naval Research can use Tech Bridges to access new ideas, technologies, and companies; to teach small businesses how to work with the Navy; to leverage academia to work on DoN problems; and to provide sailors and marines a place to bring problems or ideas and work on them.[w]
CTP-related outcomes	SoCal Tech Bridge and Marine Corps Air Station Miramar provided a test bed for a 5G network that allows industry to conduct R&D for 5G-enabled technologies while the DoD advances military capabilities.[x] Gulf Coast Tech Bridge teamed with scientists and engineers from the Naval Surface Warfare Center, Panama City Division, to build a high-quality, low-cost ventilator in response to the COVID-19 pandemic. The USSOCOM Vulcan platform launched a "Hack-a-Vent Challenge" and received 172 responses across academia, industry, and the government. The challenge recommended five prototypes including the Positive End Expiratory Pressure (PEEP) Regulated Emergency Ventilator (PRE-Vent) from the government team.[y]

[a] NavalX, "The NavalX Mission," webpage, undated b.

[b] Staff numbers are derived from NavalX, "Our Team," webpage, undated c; information on Tech Bridges is from NavalX, "2020 Tech Bridge Annual Report," 2020a, pp. 4, 6.

[c] NavalX, 2020a, p. 5.

[d] NavalX, "Centers for Adaptive Warfighting," webpage, undated a.

[e] Interview 16, Interviewee 18, September 8, 2021.

[f] NavalX, 2020a, p. 5.

[g] Interview 16, Interviewee 18, September 8, 2021.

[h] NavalX, 2020a, p. 5.

[i] NavalX, 2020a, p. 2.

[j] Interview 16, Interviewee 18, September 8, 2021.

[k] NavalX, "Tech Bridges," webpage, undated.

[l] Megan Eckstein, "Navy Now Has 15 Tech Bridges Across U.S. and in U.K. to Tackle Fleet Problems in New Ways," *USNI News*, December 14, 2020.

[m] Interview 24, Interviewee 30, September 20, 2021.

[n] NavalX, 2020a, p. 5. On scrum, see NavalX, "Military Scrum Master Course," Center for Adaptive Warfighting, 2017.

[o] NavalX, 2020a, p. 5.

[p] NavalX, 2020a, p. 5.

[q] NavalX, 2020a, p. 30.

[r] NavalX, 2020a, p. 5.

[s] NavalX, "DON Workforce Use of a Tech Bridge," video, YouTube, June 4, 2020b.

[t] Interview 16, Interviewee 18, September 8, 2021.

[u] Interview 16, Interviewee 18, September 8, 2021.

[v] NavalX, 2020b.

[w] Megan Eckstein, "NavalX Innovation Support Office Opening 5 Regional 'Tech Bridge' Hubs," *USNI News*, September 3, 2019.

[x] NavalX, 2020a, p. 5; Interview 24, Interviewee 30, September 20, 2021.

[y] NavalX, 2020a, p. 11.

SOFWERX

Snapshot of organization	**Founded:** SOFWERX is one of the DIOs under the DEFENSEWERX umbrella. It was established in September 2015 under a partnership intermediary agreement (PIA) with USSOCOM and is headquartered in Tampa, Florida.

Mission and focus: SOFWERX is a nonprofit organization that seeks to solve challenging warfighter problems through ideation,[a] networking events, and rapid prototyping by collaborating with industry, research labs, academia, and government stakeholders.[b] SOFWERX strives to bring the right minds together to solve problems for the special operations community by facilitating an innovative space known as the "ecosystem."[c]

Organizational statistics: From 2015 to 2021, SOFWERX contracted $41,185,886 through 572 purchase orders and business-to-business agreements.[d] SOFWERX has 21 employees, including three program managers.[e]

How it operates: SOFWERX operates through a PIA between DEFENSEWERX, which is SOFWERX's parent 501(c)(3), and the USSOCOM. The 11 PEOs and other USSOCOM stakeholders work closely with SOFWERX through a collaborative project order.[f] Collaborative project orders (CPOs) are stand-alone orders issued by USSOCOM agreements officers to fund specific projects under the PIA and include a statement of work, period of performance schedule, and a contract value. CPOs allow PEOs and other special operations stakeholders to present SOFWERX with a complex problem set to solve, allowing SOFWERX to convene end users, acquisition professionals, and industry representatives to collaborate on potential solutions. From 2015 to 2021, SOFWERX conducted 274 projects through CPOs.[g]

Primary customers: USSOCOM. |
| Link to CTP | **Identification functions performed:** Technology scouting; network development; problem identification, validation, and curation.

Development functions performed: S&T activities, RDT&E activities.

Enabling functions performed: Coordination, navigation and relationships, hard infrastructure to support development. |
| How they do their CTP functions | **Technology scouting:** SOFWERX hosts Tech Tuesdays, a forum specifically designed to identify and discover transformational technology that can be applied to address problems of interest to the special operations community. Tech Tuesdays are used to engage partners in industry and academia and allow them to pitch their ideas to USSOCOM and other interested government partners.[h] SOFWERX identifies new technology partners by conducting market research and environmental scans for external parties to bring into its ecosystem.

Problem identification, validation, and curation: SOFWERX identifies problems from a wide variety of sources including PMs, PEOs, warfighters, and USSOCOM headquarters directorates; USSOCOM PEOs bring warfighter problems to SOFWERX, articulated at the unclassified level.[i] USSOCOM also brings SOFWERX research topics to the Tech Tuesdays engagements, and SOFWERX searches its ecosystem to identify if a company is working to solve those problems.[j] After SOFWERX receives those problems, it conducts stakeholder mapping, writes up case studies or vignettes, assesses capability gaps, and examines how the technology will be used, with the ultimate goal of creating a strong problem statement to move forward.[k]

Network development: SOFWERX facilitates a digital network that focuses on outreach and communication for technology companies. All parties interested in joining the SOFWERX ecosystem can sign up to receive emails about ways to engage and participate with other members of the network.[l] Participants can also pass along the information from those emails to those with specific expertise. |

S&T and RDT&E activities: SOFWERX performs basic and applied research to develop technologies at TRLs 3–6. It tests and evaluates technology, facilitates agreements to develop technologies, and can support development from a minimum viable product through rapid prototype or from modifications of COTS to suit SOCOM needs.[m] In collaboration with USSOCOM, SOFWERX hosts a Rapid Prototyping Event aimed at rapidly identifying and building low-cost prototypes for problem sets.[n] If the TRL is 6 or 7, SOFWERX can recruit operators to conduct limited tests to validate the technology.[o]

Coordination: SOFWERX notifies its ecosystem participants by email about opportunities and how they can become involved. The ecosystem is made up of industry, research laboratories, academia, and government stakeholders that are interested in solving challenging warfighter problems. When USSOCOM stakeholders provide a problem set where there is not an apparent readily available solution, SOFWERX assembles a team to facilitate and further define the problem and identify the solutions.[p]

Navigation and relationships: To enhance career development, SOFWERX "offers opportunities to enhance and expand current workforce skill sets through . . . collaboration and assessment events, leadership seminars and guest lecturers."[q] SOFWERX also supports development of the next generation of scientists through its STEM outreach program for primarily junior college and four-year college students, providing opportunities for students to gain experience and knowledge in STEM skills to help the warfighters of tomorrow; sponsoring clubs, teams, and classes; and hosting STEM events.

Hard infrastructure to support development: The SOFWERX facility has meeting space to accommodate multiple meetings and workshops ranging from groups of four to 250 people, "as well as open collaboration spaces for interactive collisions. Each space is designed for spurring innovative thought and can be reconfigured to accommodate the needs of the user. The Foundry (located within the SOFWERX facility) is a rapid prototyping workshop [with capabilities including 3D printers and plasma cutters]."[r]

CTP-related outputs	SOFWERX focuses on three different kinds of innovation: *incremental* (making improvements to existing military technology), *adjacent* (taking a technology from a different market and modifying it for the warfighter), and *transformational* (creating entirely new businesses to service new markets for game-changing capabilities).[s] CTP outputs include consignment of newly demonstrated technology to a unit and acquisition through a contract (e.g., FAR-based acquisition).
Definition of success	SOFWERX fosters an environment that seeks to increase collaboration among various organizations.[t] Success for SOFWERX is based on four metrics as defined by USSOCOM: • transition (i.e., tech development that eventually moves into a program of record) • consignment (i.e., technology that is bought with an O&M contract and stakeholders purchase locally) • knowledge transfer (i.e., a new idea/product/technique that SOFWERX is able to share with USSCOM) • validation (i.e., validating the technology and defining the coalition of the willing, the acquisition strategy, and the implementation plan).[u]
CTP-related outcomes	SOFWERX is a place where prototyping, experimentation, and collaboration occur; it allows for innovation that centers on special operations. Using warfighters' input, SOFWERX aims to create solutions that meets their needs. For example, SOFWERX—in collaboration with the Army Research Lab, the Army Asymmetric Warfare Group, U.S. Army SOCOM, and Hendrick Motorsports—conducted a combat evaluation of a novel maneuver concept based on the Silent Tactical Energy Enhanced Dismount (STEED). The STEED is "a lightweight, electrically powered cart that can maneuver through different terrain such as rocks, wetlands, and sand." Using the STEED, a single operator "can maneuver 300–500 pounds at ranges of about 20 nautical miles and with its low center of gravity and versatile load deck, it has the potential to maximize ground force operations in austere conditions with limited personnel."[v]

SOFWERX also contributed to the development of a platform-agnostic AI/machine learning data management technology called Rubicon through an SBIR with Black Cape. Rubicon provides simple user interfaces, lightweight tools for data analysis, and secure data storage. Rubicon has transitioned to SBIR phase III.[w]

[a] Ideation is a concept in design thinking that focuses on open-ended development of ideas to address a challenge or issue. During the ideation phase of design thinking, participants challenge prior assumptions but are usually discouraged from judging any ideas offered during the session in order to promote exploration and generate as many ideas as possible (Thomas Both and Nadia Roumani, "Ideation/Creation Expedition Plan: A Flight Plan for Design Exploration," Stanford d.school, webpage, undated).

[b] SOFWERX, homepage, undated d.

[c] SOFWERX, "About SOFWERX," webpage, undated a.

[d] SOFWERX, "Creating Opportunities," webpage, undated b.

[e] Interview 21, Interviewee 25, September 16, 2021.

[f] Mandy Mayfield, "Need for Speed: SOFWERX Zeros In on Rapid Acquisition," *National Defense*, May 10, 2021.

[g] SOFWERX, undated b.

[h] SOFWERX, "Tech Tuesday," webpage, undated f.

[i] Interview 21, Interviewee 25, September 16, 2021.

[j] Mayfield, 2021.

[k] Interview 27, Interviewee 34, September 24, 2021.

[l] SOFWERX, "Join the Mission," webpage, undated e.

[m] Interview 27, Interviewee 34, September 24, 2021.

[n] SOFWERX, "Events: Rapid Prototyping Event," webpage, undated c.

[o] Interview 27, Interviewee 34, September 24, 2021.

[p] Mayfield, 2021.

[q] SOFWERX, undated a.

[r] SOFWERX, undated a.

[s] SOFWERX, undated f.

[t] SOFWERX, undated a.

[u] SOFWERX, undated b.

[v] SOFWERX, "Impact Box Consignment," information obtained by authors, June 10, 2022.

[w] Black Cape, "Rubicon," webpage, undated c.

xTechSearch

Snapshot of organization	**Founded:** The Army Expeditionary Technology Search (xTechSearch) launched its first competition in May 2018 at the direction of Bruce Jette, then–Assistant Secretary of the Army for Acquisition, Logistics, and Technology (ASA(ALT)). It has since expanded into a broader Army program, known as xTech.[a]
	Mission and focus: The Army xTech Program "manages the Army's prize competitions to award [prizes] and accelerate innovative technology solutions that can help solve Army challenges."[b] The program targets nontraditional businesses to uncover transformative dual-use science and technology solutions to support the Army's current needs. xTech offers cash prizes to winning participants "in addition to education, mentorship, and networking opportunities to help integrate small businesses into the Army science and technology ecosystem."[c]

Organizational statistics: Since the program's inception, ASA(ALT) has launched more than 17 competitions with over 2,485 submissions involving 20,830 evaluations from DoD stakeholders, and more than 2,200 judges from across the Army and S&T ecosystem. The program has awarded more than $15.6 million in cash prizes and led to more than $71.6 million in contracts.[d] In FY 2022, xTechSearch's budget request was $4.938 million, up from $4.817 million in FY 2021.[e] The budget covers approximately $2 million for the xTechSearch competition, administrative support to other xTech competitions, and six staff members.

How it operates: xTech has several efforts under one umbrella: the xTechSearch competitions, topic-specific competitions, xTech Accelerator, and topic-focused competitions such as the xTechInnovation Combine Advanced Energy Storage and xTechSBIR Waveform challenges. Each competition has multiple phases that typically consist of an application round (participants submit a short-concept white paper), a virtual technology pitch, and a final in-person pitch presentation during which participants showcase their solutions. Participants are down-selected during each phase and receive non-dilutive cash prizes (which do not require the company to exchange equity shares for the funding) in addition to diverse mentorship, education, and strategic exposure opportunities allowing these businesses to gain a better understanding of the Army while gaining direct exposure to key Army and commercial stakeholders. xTech's stand-alone competitions[f] can result in a follow-on contract (e.g., an OTA) with the sponsoring organization that the winner uses to develop its proposed prototype or research.[g] Finally, xTechSearch finalists participate in the xTech Accelerator program.

The xTechSearch program is a once-yearly (formerly twice-yearly) Open Topic competition with four phases: concept white papers, technology pitches, semifinals, and finals.[h] Each challenge results in one $250,000 grand-prize winner, but other finalists benefit from the challenge as well. Participating companies also receive detailed feedback from judges, mentoring and webinar services, and network with Army SMEs and program offices.[i]

Primary customers: U.S. Army.

Link to CTP	**Identification functions performed:** xTech undertakes technology scouting, customer discovery, network development, and matching problems to potential solutions. **Development functions performed:** xTech supports further development and maturation of solution concepts, S&T activities, and technology maturation (early stage). **Adoption functions performed:** xTech assists with transition primarily as a conduit but does not itself perform core adoption functions. **Enabling functions performed:** xTech conducts coordination, funding and investment of cash prizes and potential follow-on contract awards, and navigation and relationships functions (including networking, mentoring). Interviewees noted that the sponsoring organization typically finances the cash prizes, as well as SMEs and judges, but xTech has access to ASA(ALT) as a funding base if necessary.[j]
How they do their CTP functions	**Technology scouting:** The tech scouting team refines and develops problem statements, identifies key focus areas, and identifies potential businesses that can apply to the competitions. xTech maintains a close relationship with the GSA challenge.gov and various defense innovation listservs. xTech solicits companies for its tailored competitions by posting the challenge as an RFI on challenge.gov as well as other channels. The xTech team conducts market research and posts all new competition opportunities to the organization's social media channels and maintains a large email distribution list that companies can request to join.[k]

Customer discovery: xTech identifies problems through two principal channels. First, the Open Topic xTechSearch competition allows companies to pitch technologies that address specific Army modernization priorities. Second, xTech has relationships with Army program offices or other sponsors to solicit problem statements or identify technologies of interest; in FY 2021, xTech supported eight competitions originating from various Army sponsors.[l] They included, for example, the xTechInnovation Combine Advanced Energy Storage Competition and xTechGlobal: AI Challenge. xTech leverages organizations like the Army Innovation Lab and a technology scouting team to help curate the sponsor's problem statement.

Network development: The xTech program is led and administered by ASA(ALT). Through this organizational structure, xTech maintains close coordination with Army program offices, cross-functional teams, Army SMEs, and end users to develop problem statements, select judges for xTech competitions, and match mentors with companies for the xTech Accelerator. The program also coordinates with other Army innovation organizations such as the Army Innovation Lab and technology scouting teams. Interviewees noted the challenges of effectively networking across the Army, but they attempt to overcome stovepiped programs through weekly working group meetings. They acknowledged, however, that engagement with industry and information-sharing across programs remains a challenge because of limited time and resources to support synchronization activities.[m]

As noted above (*Technology scouting*), xTech maintains a close relationship with GSA challenge.gov and related listservs and online forums to exchange information and inquire about innovation efforts; however, interviewees also noted that xTech has historically established only informal, sporadic interactions with other DIOs.[n] The xTech staff is small, which makes it harder to actively engage other DIOs; nonetheless, xTech has sought to increase its engagement to share information.[o]

Matching problems to potential solutions: xTech has several competitions throughout the year, which include Open Topic competitions like the annual xTechSearch competitions and some targeting specific Army problem statements or technologies in coordination with relevant program offices or other Army organizations. These competitions facilitate a more direct matching process between a given sponsor with a particular problem and companies proposing targeted solutions.

During competitions, xTech uses a panel of judges to assess the quality of companies' submissions, incrementally narrowing the field of submissions through a series of rounds until a group of finalists is named. Participants have the opportunity to engage with Army program offices and end users throughout the competitions, creating continuous opportunities to match with potential sponsors. This often leads to connections with program offices and end users across the Army that provide funding opportunities outside the competition.[p] In previous years, finalists have been encouraged to present their proposed solutions at the annual AUSA conference and participate in the xTech Accelerator program, fostering additional interactions with potential sponsors.

Further development and maturation of solution concept: xTech competitions typically consist of multiple phases (sometimes up to four) that include concept white papers, technology pitches, semifinals, and finals. In each phase, competition judges and SMEs provide detailed feedback to companies on the content and quality of their proposals. This allows participants to refine solution concepts at each stage of the competition to better address Army needs. The judging process is supported by a transparent evaluation and feedback platform (provided by a third party, Valid Eval) that provides participants with a "heat map" highlighting strengths and weaknesses of proposals.[q] The evaluation criteria are created specifically for each competition to accurately capture the competition's problem statements.

Through the xTech program, small businesses can connect directly with Army SMEs in their fields, including engagement with Research Laboratories and Engineering Centers, PEOs, concept developers and the requirements community, reservists (through the 75th Innovation Command), Army leadership, and specialized SMEs. These connections allow for further development of technology concepts and development opportunities through multiple channels.[r]

S&T activities: xTechSearch finalists are typically given the opportunity to demonstrate a proof-of-concept or prototype at the annual AUSA conference. Likewise, xTech's targeted sponsor competitions typically allow companies to pitch and refine proofs-of-concept and prototypes, depending on the needs of the sponsoring Army organizations.

Technology maturation (early stage): The pipeline from xTech's competitions to the xTech Accelerator program constitutes the organization's primary form of early-stage technology maturation. Finalists are eligible to participate in the accelerator and to build on their final pitches, proofs-of-concept, or prototypes while learning how to work with the Army. The xTech program connects small businesses with Army and other DoD sponsors, often leading to a range of contractual vehicles through which the companies may engage in prototyping and further technology development. xTech therefore fosters early-stage development and acts as a conduit to technology maturation, development, and prototyping.

Coordination: See above, *Network development.*

Funding and investment: xTech provides prize money to companies competing in the various phases of the competitions as they are down-selected.[s] Funding for these competitions is provided both through ASA(ALT) and sponsoring organizations, depending on the arrangement. Judges are solicited through the xTech network, as well as the sponsoring organization.

Navigation and relationships: See above, *Network development.* xTech considers the mentorship, socializing, and feedback components of its competitions and the xTech Accelerator to be core to the organization's value proposition.

CTP-related outputs	xTech interviewees noted several possible outputs, including partnership agreements with prime contractors, SBIR/STTR agreements, CRADAs, OTAs, and direct contracts with Army sponsors for follow-on work.[t] Other intermediate outputs (i.e., those outputs that do not result in a tangible product for end users) include competition feedback to companies, company-generated white papers and technical papers, and mentorship arrangements.
Definition of success	xTech's objectives are to break down barriers to working with the Army, connect with nontraditional businesses to spur innovation, accelerate technology development for the Army, and provide direct exposure to Army and commercial stakeholders to transition technology solutions into the Army. According to xTech, the organization's primary objective (and ultimate measure of success) is "a full-on contract on a roadmap to a program of record"; however, this is not yet a viable measure of success since xTechSearch is "only three years old and programs of record are extremely slow."[u] Over the next five to ten years, the organization anticipates making transition to a program of record a key benchmark for success. In the short term, however, the xTech program views success as empowering small business to do business with the Army, growing this ecosystem of companies, and increasing collaborations across the S&T and user communities.[v]
CTP-related outcomes	The xTechSearch website identifies the following metrics covering the third, fourth, and fifth competitions:[w] • 500+ Army personnel engaged in customer discovery • $14.4 million awarded DoD contracts • $48.2 million total contract value applied • 800+ person-hours of mentor/company meetings

- Mentor breakdown: 37% military, 37% commercial, 26% entrepreneur
- 76% of mentors said their advice had a significant impact on the company, and 99% of mentors were willing to participate in future FedTech
- 25+ continuing engagements with venture capital groups
- 30+ strategic discussions with U.S. industrial base leaders
- 120+ unique funding opportunities shared with xTech companies
- 30 one-on-ones with the ASA(ALT) chief technology officer
- 50+ actionable FedTech-provided collisions.

However, interviewees acknowledged that these are not particularly robust outcome measures. xTech has pointed to various success stories, including LiquidPiston, a finalist from the third xTechSearch competition and a finalist from the xTechInnovation Combine Advanced Energy Storage competition. LiquidPiston has received approximately $200,000 in total prize money and has a SBIR phase II with the Army that was awarded (at the time of our interview, the award was being processed). Another success story identified by interviewees is United Aircraft Technology, a finalist from the second xTechSearch competition that has garnered Army contracts and a partnership agreement with Bell Helicopter, a prime contractor to the Army. Compound Eye, a company developing computer visualization technology, participated in the xTech Accelerator and successfully applied for SBIR phase II funding through the xTech SBIR competition; however, a subsequent SBIR funding award still took approximately eight months to complete. (Winners of an xTech SBIR competition are invited to submit an application to receive a SBIR phase II award to develop a prototype of their technology, rather than having to compete for a SBIR phase I award first.)

---

[a] Interview 35, Interviewee 46, October 19, 2021.

[b] xTech, "xTechSearch Background," webpage, undated b.

[c] xTech, "xTechSearch Background," webpage, undated b; information also obtained from xTech officials, email exchange with authors, March 14, 2022.

[d] xTech officials, email exchange with authors, March 14, 2022.

[e] The RDT&E program element (PE) appears to fund the xTechSearch competitions, but it is not clear that it funds all operations or other xTech projects such as the xTech Accelerator. Before FY 2021, xTech activities did not have a stand-alone PE. See DoD, *Fiscal Year (FY) 2022 Budget Estimates: Army Justification Book Volume 3a of 3*, May 2021, p. 169.

[f] Interview 35, Interviewee 46, October 19, 2021.

[g] xTech officials, email exchange with authors, March 14, 2022.

[h] xTech, undated b.

[i] Interview 35, Interviewee 46, October 19, 2021.

[j] Interview 35, Interviewee 46, October 19, 2021.

[k] Interview 35, Interviewee 46, October 19, 2021.

[l] Interview 35, Interviewee 46, October 19, 2021.

[m] Interview 35, Interviewee 46, October 19, 2021.

[n] Interview 35, Interviewee 46, October 19, 2021.

[o] According to one interviewee, at one point "the OSD A&S office started a weekly sync between DoD innovation programs that was run out of the industrial policy office. That's the same group that started the trusted capital marketplace. They had a keen interest in seeing what all the innovation programs were doing, [the] companies being selected, and bring[ing] them into the trusted marketplace. . . . More recently, through the trusted capital working groups, we've talked a lot more about data sharing. [Previously], a friend at SOFWERX would call and ask, 'Hey, have you seen technologies on counter-drones?' I would email them a listing. That's not a sustainable way to do it" (Interview 35, Interviewee 46, October 19, 2021).

[p] Interview 35, Interviewee 46, October 19, 2021.

[q] Interview 35, Interviewee 46, October 19, 2021.

[r] xTech officials, email exchange with authors, March 14, 2022.

[s] Interview 35, Interviewee 46, October 19, 2021.

[t] Interview 35, Interviewee 46, October 19, 2021.

[u] Interview 35, Interviewee 46, October 19, 2021.

[v] Interview 35, Interviewee 46, October 19, 2021. Interviewees also noted that they judge success by whether learning occurred. That is, if the businesses "learned something about their technologies, what to do different, whether to pursue Army, if they had access and got in front of the right expert, that's a success."

[w] xTech, "Accelerator," webpage, undated a.

Technology Case Studies

Black Cape

Technology Story

What Is the Technology?

Black Cape develops platform-agnostic data analytics, machine learning (ML), and artificial intelligence (AI) applications and services to aid clients in addressing mission-specific needs. Its products include a geospatial analysis collaboration platform; ML-enabled geospatial analysis; and a data ingest, storage, and analysis platform that can support users in tactical environments, where low bandwidth and power are constraints, as well as at more centralized locations.[1]

Black Cape develops technology in three focus areas. The first is mission applications tailored to customer needs that are intended to be simple and intuitive, secure (for sensitive data), and accessible through multiple devices. The second focus area is ML and AI algorithms. The third focus area is data analytics, where the company's data scientists and engineers work with customers to conduct tailored analysis.[2] Black Cape's data ingest, storage, and analysis suite of tools called Rubicon was partially developed as part of an SBIR project in collaboration with the U.S. Special Operations Command (USSOCOM).

What Is the Company, and Who Are the Founders?

Black Cape is a veteran-owned small business with offices in Arlington, Virginia, and Austin, Texas; it emerged from the "stealth mode" temporary state of secrecy as an early start-up in December 2019.[3] The business was started by three "serial entrepreneurs": Abe

---

[1] Black Cape, "Products," website, undated b.

[2] Black Cape, "Focus Areas," website, undated a.

[3] Stealth mode is used to develop a product or service or to develop technology in complete secrecy before publicizing the company or its offerings. Stealth mode can cover the company as a whole or certain products within the company. See Black Cape, "Black Cape, Inc., Announces Emergence from Stealth Mode," press release, December 18, 2019.

Usher, a former Army infantry officer, was a software engineer after retiring at multiple companies, including SAIC and Google. He embarked on his entrepreneurial career with HumanGeo, a data analytics company that was subsequently acquired by the Radiant Group. Radiant was then acquired by DigitalGlobe, a commercial imagery analysis company, where Usher became chief technology officer. LTC (ret.) Al Di Leonardo is a former special operations officer who developed a new intelligence analysis cell while at SOCOM.[4] This intelligence analysis cell combined software engineers, intelligence analysts, and experts in the human terrain to accelerate analysis identifying targets.[5] LTC (ret.) Brian Poe is a former U.S. Army Intelligence Officer who previously led unique intelligence capabilities across the Army and the intelligence community. Other company employees include Sam Stowers, Black Cape's head of product development, and Scott Fairgrieve, vice president of user experience. Stowers spent many years in SOCOM and formerly led innovation efforts at a subordinate SOCOM unit. Stowers additionally spent time at the Defense Digital Service (DDS), the DoD's "SWAT Team of Technologists" charged with innovation in both technology and acquisition processes. Fairgrieve has spent over 20 years working in national security technology, including at Northrop Grumman and HumanGeo. Di Leonardo, Poe, and Usher founded Black Cape with the idea of targeting both the public and private sector as customers. In the public sector, Black Cape has focused on the national security agencies, including the intelligence community and the DoD. Black Cape also has a growing commercial customer base.

How Was the Technology Developed?

Black Cape originally developed unique intellectual property (IP) for private-sector businesses to process large volumes of geospatio-temporal data and to use a knowledge graph to understand large volumes of text data.[6] Black Cape combined these two offerings and successfully proposed it to a SOCOM SBIR phase I opportunity through SOFWERX. During the SBIR phase I feasibility study, Black Cape evaluated a variety of SOCOM mission data to identify challenges in data storage and processing. The combined efforts of this work culminated in the development of the Rubicon platform in a SBIR phase II. Rubicon satisfied SOCOM's requirements for platform-agnostic data storage, data processing, and the delivery of enterprise analytic services to special operations forces. Black Cape's research has led to the development of new IP and patentable technology and has resulted in multiple follow-on contracts across a range of contraction mechanisms (e.g., extended SBIR phase IIs, multiple SBIR phase IIIs).

---

4 Eli Gorski, "HumanGeo Acquired by Radiant Group," *FCW*, July 27, 2015.

5 Black Cape representative (Interview 41, Interviewee 53), November 10, 2021.

6 A knowledge graph is a method of illustrating the relationships between objects, entities, or nodes of a network. It is also known as a semantic network.

DoD Story

How Did They Get Introduced to DoD?

The founders had experience coming from the special operations community and wanted to contribute back to the community. They recognized, however, that SOCOM is a unique organization to engage successfully as a small business and would also require significant investments to build a team capable of delivering value to a discerning customer. "Few companies work with SOCOM because of the challenging mission set and other programmatic limitations. Tampa, Florida, does not have a lot of local tech activity like Washington, D.C., or Silicon Valley. Working with SOCOM is the opportunity to have impact, but SOCOM has the same limitations in acquisition that the rest of the public sector has, only with the unique benefit of highly motivated personnel. . . . The whole public sector is challenging, but SOCOM is worth the investment."[7] Black Cape has pursued work with the Air Force and the other services and has successfully won Air Force SBIRs related to additional modifications of the Rubicon platform.

Interaction with Innovation Ecosystem

Black Cape has received SBIR phase I and II awards from SOFWERX, SOCOM's innovation organization with special authorities to manage SBIRs and research and development agreements (RDAs). The phase I award in 2020 was $144,000 for Black Cape's STARbase (Spatial-Temporal Analysis and Reasoning Database Engine) platform-agnostic data storage infrastructure (PADSI) technology.[8] The phase II award was $744,813 in August 2020 to extend development of the PADSI technology.[9] Sam Stowers led Black Cape's SBIR-funded project in conjunction with Scott Fairgrieve.[10] Together, they were able to combine their extensive mission, acquisitions, and technology experience into a successful SBIR and build a compelling technology platform.

What Is the Potential Military Application and Market?

DoD has placed significant emphasis on data analytics, machine learning, and AI to augment human decisionmaking and support autonomous systems. DoD established the Joint Artificial Intelligence Center (JAIC) in 2018 to "accelerate the delivery of AI-enabled capabilities, scale the Department-wide impact of AI, and synchronize DoD AI activities to expand Joint Force advantages."[11] The DoD AI strategy emphasized working with commercial and academic partners to develop AI capabilities. As more data is produced through deployed

---

[7]   Black Cape information to authors, June 29, 2022.

[8]   SBIR.gov, "STARbase: An Innovative Solution for Platform Agnostic Data Storage," webpage, undated b.

[9]   SBIR.gov, "Platform Agnostic Data Storage Infrastructure," webpage, undated a.

[10] Black Cape information to authors, June 29, 2022.

[11] U.S. Department of Defense, *Summary of the 2018 Department of Defense Artificial Intelligence Strategy*, Washington D.C., February 12, 2019.

sensors, DoD is exploring how it can make better use of the data it has, including how to analyze and act on data in contested environments (the so-called tactical edge) where bandwidth is constrained and reach-back to centralized analytic capacity may be disrupted. At the same time, DoD wants to improve the use of the data it has in these centralized repositories to improve strategic and operational-level analysis. One of the challenges in AI is the problem of "black box" algorithms—processes for evaluating data and producing results that are not transparent or explainable to the human user or consumer.[12] The demand for explainable AI is high in DoD because decisionmakers want to understand how an AI algorithm arrived at the analytic result. Practical applications in defense contexts range from improved discrimination of targets in full-motion video feeds from unmanned aerial vehicles (UAVs) to better predictive analytics on maintenance of aircraft and other systems. Black Cape's technology is positioned to address these types of analytic needs in a straightforward and transparent way that answers specific mission needs. Areas of interest are target discrimination; processing, exploitation, and dissemination (PED); and the application of data analytics through the operations and intelligence processes.[13]

Status: Are They Selling to DoD?

Black Cape has entered into a SBIR phase III contract with SOCOM. This phase III contract allows broad government purchasing of Black Cape products, services, software, and data, including the ability to purchase the entire Rubicon suite of technologies.[14] Black Cape's additional SBIR awards from the Air Force and Army are enabling further R&D and may eventually translate to additional phase III awards.

Commercial Marketplace Story

What Is the Potential Commercial Application and Market?

Data analytics is a large and growing market, estimated to be worth more than $200 billion worldwide in 2020 and projected to grow to more than $500 billion by 2028.[15] The market has several large corporations that are significant players, including Microsoft, Salesforce, and Experian. These large businesses are more likely to provide data analytics platforms and services that are marketed to a broad array of industries and markets, with less focus on tailored solutions. Black Cape's focus on mission-tailored analytics may provide it opportu-

---

[12] Cynthia Rudin and Joanna Radin, "Why Are We Using Black Box Models in AI When We Don't Need To? A Lesson from an Explainable AI Competition," *Harvard Data Science Review*, Vol. 1, No. 2, November 22, 2019; Ariel Bleicher, "Demystifying the Black Box That Is AI," *Scientific American*, August 9, 2017.

[13] Black Cape information to authors, June 29, 2022.

[14] Black Cape information to authors, June 29, 2022.

[15] Fortune Business Insights, "Big Data Analytics Market: 2021 Size, Growth Insights, Share, COVID-19 Impact, Emerging Technologies, Key Players, Competitive Landscape, Regional and Global Forecast to 2028," *GlobeNewswire*, December 16, 2021.

nities to target its services to small and medium-sized businesses that need tailored solutions but do not have the workforce to do it themselves. STARBase has also been used by multiple Fortune 500 companies to process large volumes of geospatio-temporal data. Black Cape has partnered with several well-known technology companies including Carahsoft, Cloudera, and Splunk to integrate its technology into their services and products.[16]

Status: Are They Selling on the Commercial Market?

Black Cape has customers in both the public sector and private sector. Black Cape is currently focused on expanding its work in the public sector. Fortune 500 companies have used Black Cape products to process data for financial services and other highly regulated industries.

Challenges

What Challenges Did They Face in Engaging with and Selling to DoD?

Black Cape notes that programs like SBIR are challenging to navigate. "There is only so much transparency from the SBIR processes; [applying for an SBIR] was still a shot in the dark. It is a leap of faith."[17] Black Cape has faced challenges gaining traction with other services through the SBIR process, indicating that serial SBIR performers appear to capture the bulk of SBIR awards. "It is hard to penetrate [the military services]; they get an insane amount of SBIRs. There is still an ecosystem around these customers. Most of the SBIR continue to be awarded to the usual suspects."[18] This indicates that small businesses see SBIR as a helpful funding source, but it is often a semiclosed system where previous performers dominate.

Black Cape also notes that new entrants face an uphill struggle against traditional DoD contractors that can sell "internal innovation" using existing software or system integrators that do not require the involvement of new entrants. For example, Black Cape's focus on geospatial analysis puts it in competition with companies like Maxar that have been in the defense industry for decades and can persuade DoD to pursue integrated technology solutions that appear easier to buy from one company rather than having to bring together multiple companies to address an issue.[19] It is also challenging to transition early-stage technologies that have been awarded SBIR funding into programs. Major AI programs have historically purchased engineering services (e.g., software engineering labor) rather than products or have selected a limited number of more well-known products like Palantir in work at the JAIC. Major programs such as the Joint All-Domain Command and Control (JADC2) are seeing few, if any, SBIR technologies mature into the program. Black Cape notes that small companies are producing a lot of innovation in AI and machine learning, but

---

[16] Black Cape information to authors, June 30, 2022.

[17] Interview 41, Interviewee 53, November 10, 2021.

[18] Interview 41, Interviewee 53, November 10, 2021.

[19] Interview 41, Interviewee 53, November 10, 2021.

these technologies stall at higher levels of technology readiness (TRL 6 and above) because customers do not have the "color of money" to pay for development, prototyping, and testing or, conversely, the right type of money to transition these technologies into programs of record.

Black Cape also noted that program managers and other contracting officials in organizations like SOCOM have had to adopt new authorities like the Partnership Intermediary Authority (the mechanism used to support contracts through SOFWERX) or learn how to use SBIRs. Proper use of these authorities takes time to master. In other services or organizations, Black Cape has lamented that SBIRs are sometimes treated like standard FAR-based contracts with "free money" (i.e., a grant with no expectation of a return to the awarding agency) instead of the innovation grants they were intended to be, which leads to slower contracting (by recompeting as a FAR-based contract after the successful SBIR). Black Cape also said there is often a lack of a clear plan for transitioning to a program of record as technologies emerge from the SBIR programs.[20] According to Black Cape, SOCOM's vision for transition of an SBIR into a program of record was clear and ultimately successful—not every SBIR program is capable of the same vision.

How Did They Address These Challenges?

Black Cape's SOCOM SBIR was transitioned to a FAR-based phase III contract at the completion of its phase II extension in the summer of 2021. Based on the success of Black Cape's first phase III, SOCOM launched a follow-on phase III that resulted in a $49 million indefinite delivery/indefinite quantity (IDIQ) contract.[21] The company continues to pursue SBIR awards from the services and is making a pitch for its technology playing a role as a common data layer (middleware) to integrate emerging SBIR technologies. The company is committed to public-sector customers, given the founders' backgrounds, but they also recognize that commercial contracts are important for the long-term sustainability of the company.

What Is Next for the Company?

While continuing to pursue defense customers, Black Cape is developing more business-to-business relationships to have its technology available to a larger number of customers through larger platforms (e.g., through Rubicon's shared data layer offering). On the government sector side, Black Cape is also exploring ways to subcontract its software services to larger defense businesses, despite the fact that it can be difficult to protect intellectual property in these relationships.

---

[20] Interview 41, Interviewee 53, November 10, 2021.

[21] Black Cape information to authors, June 30, 2022.

Compound Eye

Technology Story

What Is the Technology?

Compound Eye is developing computer vision technology called VIDAS (Visual Inertial Distributed Aperture System) that uses sophisticated software and neural networks to simultaneously reconstruct the surroundings of a robot or vehicle and classify everything in the environment using data collected from off-the-shelf cameras modified by Compound Eye and attached to that robot or vehicle.[22]

VIDAS requires both parallax and semantic cues. Parallax allows depth sensing by using two or more perspectives from cameras to triangulate the distance of any object. According to Compound Eye's website, "with a picture, some prior knowledge and experience, it's possible to solve a 3D scene." Humans use semantic cues that are learned over time to help solve scenes. If a person knows the average size of an object in a picture, they can approximate the size of other objects. Humans learn that buildings and trees are typically perpendicular to the ground, closer objects seem larger than farther objects, and an object that obstructs the view of another object must be closer to them. Computers can also learn these semantic cues.

Compound Eye's technology combines parallax and semantic cues. It has invented techniques for parallax-based depth sensing using mounted cameras and (as described on the company website) "self-supervised approaches to training neural networks for monocular depth estimation and new ways to calibrate cameras online so that robots can operate indefinitely without adjustment—in real time on embedded hardware."

In simple terms, Compound Eye captures a scene using two or more cameras, determines the distance to every point in the scene using parallax and semantic clues, then uses an integrated single framework to provide accurate depth sensing of every pixel, all in real time.

What Is the Company, and Who Are the Founders?

Jason Devitt cofounded Compound Eye (his third venture-backed start-up) in 2015 and is the CEO. He became interested in computer vision for mobile phones and depth sensing in 2014. His thesis at that time was that each new class of sensors on mobile phones opened markets and created new opportunities valued in the billions of dollars and that computer vision could do the same. Compound Eye is a privately held computer software company headquartered in Palo Alto, California, with approximately 20 employees. Employees have come from the United States, Ireland, Taiwan, Russia, and India; they have worked for Google, Apple, Amazon, and Oculus and founded six companies. Compound Eye's mission is to "bring superhuman senses to existing vehicles and to enable mass production of fully autonomous cars and other robots."[23]

---

[22] xTech, "xTechSearch 5 Finalist Technology Overview: Compound Eye," video, YouTube, September 16, 2021.

[23] LinkedIn, Compound Eye, undated b.

How Was the Technology Developed?

As previously stated, Compound Eye's thesis was that a new sensor, in this case computer vision and depth estimation on phones, could open markets and opportunities worth billions of dollars. Compound Eye showed its technology to individuals in the commercial market, and it showed promise. Eventually, Compound Eye was introduced to a major robotics company that became an anchor customer and led to a first round of venture capital.

What Are the Potential Military and Commercial Applications?

VIDAS is a deep technology, which is a generic term for enabling technologies that are not focused on specific end-user services, such as AI and quantum computing.[24] Compound Eye believes that there are three primary dual-use applications for VIDAS: superhuman vision, advanced driver assistance systems (ADAS), and fully autonomous robots.

Superhuman vision includes applications such as telepresence. Using augmented reality (AR) technology such as the Army's Integrated Visual Augmentation System (IVAS), VIDAS could allow the broadcast of a three-dimensional (3D) view of a scene to an incident commander 1,000 miles away. Another application of superhuman vision would be enabling a soldier inside a windowless vehicle to see through the walls of that vehicle for situational awareness of the environment. VIDAS could also allow soldiers to record the data from a drive and create a 3D map of the route in real time.[25] A military application with limited commercial interest at this time is the ability to drive in complete darkness.[26]

ADAS includes semiautonomous features in a vehicle such as emergency braking and lane keeping. For the military, these commercial applications also apply. ADAS features using VIDAS can reduce the cognitive load of driving, which allows soldiers to focus on other activities in and around the vehicle. VIDAS can also reduce the skill necessary to drive military vehicles, which decreases the necessary training time for soldiers. Overall, this technology can make vehicles safer.[27]

The final application of VIDAS is enabling robots to see the world as humans see it, enabling fully autonomous robot or unmanned vehicle operation.[28]

DoD Story

How Did They Get Introduced to DoD?

Compound Eye received a cold email from Eagle Point Funding, an Israeli company that helps companies win government grants and contracts, stating that Compound Eye may be

---

[24] Ingrid Lunden, "What Do We Mean When We Talk About Deep Tech?" *TechCrunch*, March 11, 2020.

[25] xTech, 2021.

[26] Interview 42, Interviewee 54, November 10, 2021.

[27] xTech, 2021.

[28] Interview 42, Interviewee 54, November 10, 2021.

eligible for a non-dilutive funding opportunity. Skeptical at first, Compound Eye looked into it and learned about SBIRs. One of the programs that Eagle Point introduced Compound Eye to was the U.S. Army's xTechSearch. Compound Eye sent white papers to various federal government organizations to explain its technology and seek funding opportunities. Until this point, Compound Eye had no prior contact with the Army or DoD. According to Compound Eye, as a deep-tech or frontier-tech company with an enabling technology, it is not always obvious what the product will be. Compound Eye became interested in the DoD because the DoD is funding deep technologies.

Interaction with Innovation Ecosystem

Compound Eye first participated with DoD and DIO through xTechSearch, "a competition sponsored by the ASA(ALT), targeting small businesses to uncover novel dual-use science and technology solutions to tackle the Army's most critical modernization challenges."[29] Compound Eye decided to enter xTechSearch because there were low barriers to entry. Compound Eye had to submit a white paper and pitch its technology to a panel of judges over successive rounds. The xTechSearch program filtered the initial 400 to 500 white papers down to ten companies that participated in an in-person event. After xTechSearch down-selected its white paper, Compound Eye participated in three rounds of pitches to a panel of judges with increasingly focused questions on the technology and higher-ranking individuals. Compound Eye found the feedback from the judges on their technology and applications incredibly useful for refining potential product and use cases. Another aspect of the program is the xTech accelerator managed by FedTech.[30]

The xTech accelerator included a series of teleconferences and lessons on various aspects of the Army, how the SBIR program works, and how to do business with the Army, which was a useful educational program for Compound Eye. The accelerator ran concurrently with preparing the participants for successive rounds of xTechSearch, which were complementary in Compound Eye's view.[31]

Compound Eye also had the opportunity to engage with Army end users and warfighters. Through these conversations, the company learned about the importance of passive-sensing night vision and how the end users value mass-produced products like smartphones and cameras.[32] After these engagements, Compound Eye pivoted from its initial white paper topic, mapmaking, to passive-sensing applications.

---

[29] xTech, "xTechSearch Background," webpage, undated b.

[30] Interview 42, Interviewee 54, November 10, 2021. The xTech accelerator helps xTechSearch finalists achieve success through mentorship, educational programming, venture-building activities, and strategic exposure: to ultimately transition their technologies into the Army and thrive as small businesses. The accelerator operates nationally, 90 percent virtually, and is open to all technology verticals." FedTech, "Accelerators," webpage, undated.

[31] Interview 42, Interviewee 54, November 10, 2021.

[32] xTech, 2021.

Compound Eye also used the knowledge and experience gained from xTechSearch to apply to a separate xTech program called xTech SBIR for a cash prize and direct to phase II SBIR contract. In November 2020, Compound Eye submitted a 15-page slide presentation, which made the process less burdensome than a 50-page proposal, and was notified in January 2021 that it would receive a contract award.[33] The Army promised that the company would be notified in 60 to 90 days and met that milestone; however, it took approximately eight months for Compound Eye to receive the contract award to further mature the technology, which it did on September 30, 2021. Being mindful of the chasm between phase II SBIRs and getting to a program of record, Compound Eye's goal is to build a product, not a prototype, during the SBIR that can be adopted immediately.[34]

What Is the Potential Military Application and Market?

Compound Eye originally entered xTechSearch because it thought the government might be interested in mapping and geospatial applications using 3D sensing to construct the environment.[35] The Army Corps of Engineers (ACE) performs these tasks for mapmaking using light detection and ranging (lidar). Compound Eye proposed VIDAS as a lower-cost method than lidar to perform these tasks.

In addition to the use cases described above, VIDAS has a key feature for DoD problems: It is 100 percent passive detection with zero signature. The Army currently uses active sensors—the same technology as used in the commercial sector—for night vision, lidar, or radar. There are methods of obfuscating radar signatures, but not with lidar. Lidar uses lasers, which according to Compound Eye, "lights you up like a Christmas tree,"[36] so adversaries can see and detect you coming. VIDAS uses cameras, so there is no signature, and Compound Eye believes this is a critical enabling technology for autonomous vehicles and robots in the DoD.

Commercial Marketplace Story

What Is the Potential Commercial Application and Market?

The potential commercial market and problem sets include superhuman vision, ADAS, and fully autonomous robots. Compound Eye drove on every street in San Francisco a single time in a car equipped with its prototype and was able to construct a 3D model of the entire city. The commercial market for this type of vision and mapmaking is valued at more than $20 billion currently and accounts for approximately two-thirds of all lidar sensors sold globally. This opportunity is Compound Eye's near-term focus. ADAS is already a

---

[33] Interview 42, Interviewee 54, November 10, 2021.

[34] Interview 42, Interviewee 54, November 10, 2021.

[35] Interview 42, Interviewee 54, November 10, 2021.

[36] Interview 42, Interviewee 54, November 10, 2021.

$25 billion market, and VIDAS can be an enabling technology for it.[37] Finally, fully autonomous robots is a nonexistent market today but will be the largest market of all, in Compound Eye's estimation. The trucking industry faces a critical shortage of skilled drivers and is worth $800 billion today; creating the visual system for robots and vehicles that can be used in applications such as trucking is Compound Eye's ultimate goal.[38]

Status: Are They Selling on the Commercial Market?

Compound Eye currently receives material revenue from a confidential customer and has raised $8 million in venture capital.[39]

Challenges

What Challenges Did They Face in Engaging with and Selling to DoD?

The Army primarily conducted xTechSearch virtually because of COVID-19, which made it difficult for Compound Eye to build networks and gain informal knowledge about priorities and problems. FedTech provided introductions to key stakeholders in the Army, which allowed Compound Eye to receive valuable feedback in one-on-one sessions; however, Compound Eye believes that judging events are more valuable in person and that holding most of xTechSearch virtually was a major limitation imposed by the pandemic. Compound Eye found the in-person final round to be beneficial and easier to share information between the company and stakeholders.

Overall, it has been difficult for Compound Eye to pitch and receive feedback on its technology outside of xTechSearch. Compound Eye faced challenges with meeting decision-makers: not only people and organizations with requirements and funding, but also Army researchers to exchange ideas with, help better understand problems, and discover which components of its technology are most promising. The company also had issues finding consistent technical points of contact (TPOCs) to evaluate its technology in conjunction with SBIR applications, a required step to enable the company to receive SBIR contracts.

As mentioned earlier, Compound Eye entered xTechSearch because there were low barriers to entry: The company submitted a white paper and then pitched its technology and applications to a panel. This process is similar to how Compound Eye normally conducts business, rather than having to complete voluminous proposals typical in government acquisitions. However, there were challenges: Compound Eye also found it difficult to identify DoD funding opportunities, learn how to do business with DoD, identify prospective end users and potential customers to explore use cases, and understand the acquisition process and how to navigate those processes.

---

[37] xTech, 2021.

[38] xTech, 2021.

[39] Crunchbase, "Compound Eye, Financials," webpage, undated.

How Did They Address These Challenges?

Most of the acute challenges regarding xTechSearch are in the past, but Compound Eye provided some suggestions and lessons for improvement, which are discussed below.

What Is Next for the Company?

Compound Eye is currently trying to expand its customer base to include the Air Force. Compound Eye's near-term priorities related to the DoD are to fulfill its xTechSearch SBIR phase II contract and further develop the technology and mature its product. Compound Eye is also seeking to develop CRADAs with Army Research Laboratories (or other laboratories) that work on robotic and autonomous systems. Compound Eye may seek to partner with the U.S. Army Combat Capabilities Development Command/Ground Vehicle Systems Center (DEVCOM/GVSC)[40] and night-vision lab on a one or more CRADAs.

Compound Eye currently has commercial revenue and is interested in refining its technology so that it meets the needs of the widest number of possible applications and highest number of end users. In an interview with the authors, the company explained that it is preparing a submission for AFWERX's next Open Topic and hopes to make inroads with the Air Force through an AFWERX SBIR. Additionally, Compound Eye wants to build relationships with prime contractors because it recognizes that VIDAS will be a component technology, and it would prefer to be a subcontractor in large programs.[41]

With respect to doing business with the Air Force, Compound Eye sees AFWERX as the "front door" to the Air Force and plans on submitting a proposal for the next AFWERX topic. The Air Force has applications for VIDAS, but Compound Eye is still trying to discover what the biggest need is for the Air Force and how to do business with them. Compound Eye used a capture company with a previous AFWERX submission but did not win a contract award. According to Compound Eye, the company needs to talk to customers and read between the lines to find the real needs. Learning to do this with the Army took time, and it needs to do the same with the Air Force.

Lessons for Defense Innovation Organizations

Compound Eye suggested that DIOs should have more of a presence in Silicon Valley in order to perform more active technology scouting, go out and meet with start-ups, and figure out what is happening and who is doing what. For a very long time, the Army and Air Force operated under the assumption that if there was a good opportunity, it would come to them, but both have learned that that is no longer the case. To illustrate this point, automotive companies based in Michigan have a large presence in Silicon Valley. Activities such as networking, attending events, and communicating with the start-up community that they are open for business can go a long way; with that in mind, Compound Eye noted that most entrepreneurs do not need handholding from the DoD—rather, the DoD should point

---

[40] U.S. Army, "DEVCOM Ground Vehicle System Center," webpage, undated.

[41] Interview 42, Interviewee 54, November 10, 2021.

these entrepreneurs in the right direction and communicate its priorities, then entrepreneurs can figure it out.[42] This is difficult to do, however, if the DoD does not have a presence where these companies reside. More generally, Compound Eye suggested that DoD needs to offer companies more low-friction ways for companies to identify opportunities and talk to potential customers.

Additionally, Compound Eye suggested that programs such as xTechSearch are relatively cheap accelerators and need to be marketed more aggressively. There are some start-ups that are not paying attention to the DoD, but other start-ups might be interested in prize challenges or non-dilutive funding opportunities—for example, where they could submit a white paper and present an idea for a chance to receive $10,000—if they were aware of the opportunities.[43]

Given the increase in SBIR opportunities marketed at nontraditional companies and numerous funding opportunities, Compound Eye notes that more capture companies—companies that help small businesses and other companies receive non-dilutive funding from the government—have emerged to help start-ups, many of which are particularly focused on grants and SBIR contracts. Small businesses looking for funding opportunities that are unfamiliar with or need help doing business with the government typically pay these capture companies a retainer fee or a percentage of contract awards in exchange for assistance with writing proposals and winning contracts. It appears to be a fee-for-service market specific to the defense (or government) small business innovation space.

---

[42] Interview 42, Interviewee 54, November 10, 2021.

[43] Interview 42, Interviewee 54, November 10, 2021.

Distributed Spectrum

Technology Story

What Is the Technology?

Distributed Spectrum LLC has conceptualized and developed a modular and scalable hardware-agnostic radio frequency (RF) spectrum monitoring system that can detect, classify, and localize RF signals in a variety of complex environments. The system uses commodity-grade off-the-shelf sensor hardware coupled with proprietary software and signaling processing approaches to analyze the RF environment in real time at low cost.[44] Distributed Spectrum's technology has a modular hardware and software design that is scalable, flexible, and can integrate with existing networks. Its algorithms can detect and classify signals in real time, on the sensor itself, without transmitting data back to another processor.[45] The sensors can act in "stand-alone mode," or sensors can be networked together across an area; the algorithms can detect and classify signals in real time, on the sensor itself, without transmitting data back to another processor.[46]

What Is the Company, and Who Are the Founders?

Distributed Spectrum was founded in August 2020 by three Harvard students. The team had previously worked with each other on class projects at Harvard, where they are pursuing technical bachelor's degrees (in electrical engineering, physics, and computer science).[47] Each of the founders had prior experience working for large commercial and defense companies, including Microsoft, Google, Lockheed Martin, and Raytheon. During their time at Harvard, they had considered collaborating to develop a system that integrates low-cost commercial sensors with novel software to help industries detect interference within the RF spectrum, specifically in complicated, dense RF environments. When Harvard announced COVID-related restrictions on the numbers of students permitted on campus for the 2020–2021 school year, the students decided to take a year off to start a company and to focus on developing the RF monitoring product that they had conceived. The three-person team founded Distributed Spectrum LLC in August 2020.

How Was the Technology Developed?

In fall 2020, the founders began to develop their technology. Over a period of a year, they developed the control software and algorithms necessary to detect, classify, and localize

---

[44] Distributed Spectrum LLC, homepage, undated; Distributed Spectrum LLC, "Product Brief," on file with authors, 2020.

[45] Distributed Spectrum LLC, "Product Brief," on file with authors, 2020; Air Force Research Laboratory, "Tactical Assault Kit," webpage, undated.

[46] Distributed Spectrum LLC, "Product Brief," on file with authors, 2020.

[47] Interview 20, September 15, 2021.

RF transmitters with low-cost commodity hardware. The Distributed Spectrum team purchased small quantities of off-the-shelf computing devices and radio hardware with their own funds. Using this hardware, they were able to evaluate technical concepts foundational to the product they eventually wanted to develop. They installed a variety of $50 sensors around Cambridge and Somerville, Massachusetts, and used these sensors to test signal detection and localization capabilities. By February 2021, Distributed Spectrum had developed a minimum viable product capable of demonstrating spectrum situational awareness capabilities.

DoD Story

How Did They Get Introduced to DoD?

In fall 2021, as the founders started to build their technology, they applied for a National Science Foundation (NSF) SBIR phase I in December 2020, seeking $235,000 to further develop their low-cost network of sensors and evaluate its capabilities on a broader scale. While waiting to hear back from NSF on their application, the team began searching online for opportunities to work with DoD.

Interaction with Innovation Ecosystem

During their internet search, the Distributed Spectrum team came across the NSIN website and discovered that NSIN was seeking applicants for a "hackathon" event, an opportunity for students, academics, entrepreneurs, start-ups, engineers, and others to address a challenge that is relevant to both the defense and commercial markets.[48] NSIN solicits and convenes participants over a weekend to develop ideas to respond to the identified challenge, and the hackathon participants work with NSIN-identified mentors and subject-matter experts to develop their solution concept.[49] Hackathon participants are not required to build a prototype. The teams/individuals present their novel solution concept to a panel of judges in a virtual "pitch." Distributed Spectrum was attracted to the low barrier to entry of the competition—the online application required only a small amount of time for the team—and minimal investment of time required to participate in the hackathon competition. They applied and were accepted to participate in the "Mad Hacks: Fury Code" hackathon.

In all, 500 registered participants, including the Distributed Spectrum team, joined this hackathon from February 5 to 19, 2021; the aim was described as follows:

> [Participants seek to] develop technologies or systems to help human-controlled and autonomous vehicles operate through cyber-attacks or other instances of electronic warfare. This is a problem because human-controlled and autonomous vehicles play critical

---

[48] Interview 20, September 15, 2021; National Security Innovation Network, "Hackathons Redefined," webpage, undated c.

[49] NSIN, undated c.

roles in mobility and survivability, but [they] are vulnerable to cyber-attacks that could disable the system and leave troops unable to communicate and coordinate. Given the diversity of the vehicles used, along with the adversary force capabilities, successful cyber-attacks are all but inevitable.[50]

Distributed Spectrum was one of nine teams selected from the 500 original participants to move on to the second phase of the hackathon, a pitch event where each team gave a presentation on their technology solution concept.[51] At the pitch event, Distributed Spectrum presented to prospective DoD customers including the 1st Cavalry Division, U.S. Army, U.S. Army DEVCOM Ground Vehicle Systems Center, and Next-Generation Combat Vehicles Cross-Functional Team. Together with Dell Technologies, these prospective DoD customers evaluated the technology pitches, assessing whether the concept would meet their end-user needs and solve the identified problem, and selected winners.[52]

The team presented their RF Threat Detection System solution concept at the pitch event on February 26, 2021. The judges selected four winners, and the Distributed Spectrum team won the grand prize, a $25,000 contract to help the winner continue developing their concept.[53] Through the hackathon, Distributed Spectrum discovered that DoD did not have a modular and scalable system that would allow warfighters to characterize signals in the RF environment across a range of platforms. On the basis of this information, Distributed Spectrum reshaped and broadened its technology concept, which the team members had originally envisioned as a system to identify interfering signals. As they learned more about DoD end-user needs during the hackathon, the team modified and broadened their technological solution: Rather than focusing on interference, they proposed a general-purpose system that could perform signal detection, classification, and localization capabilities across a range of hardware platforms and deliver interoperable outputs to DoD information systems.[54] In the ground-vehicle context, this system could help a military user determine if a failure of vehicle communications or navigation devices is the result of a threat (jamming) or an internal malfunction.

After the hackathon, NSIN tailored the grand prize of a $25,000 six-month contract specifically to Distributed Spectrum to support customer discovery and development of a dual-use technology. As part of the contract, Distributed Spectrum worked with an NSIN mentor who provided navigation and coaching and introductions to prospective end users, so Distributed Spectrum could further develop military use cases. For example, the mentor taught the team to "go through the green book and look at the 6.1–6.4 budget. Look for stuff, project

---

[50] National Security Innovation Network, "Mad Hacks: Fury Code Concludes with Four Winners," February 26, 2021c.

[51] NSIN, 2021c.

[52] NSIN, 2021c.

[53] NSIN, 2021c.

[54] NSIN, 2021c; additional NSIN information to authors, April 4, 2022.

codes, money that aligns with their technology. They have an RF spectrum analysis tool; who is working on that? Where does the money go? To which program office do the funds go? By doing this, they can reverse-engineer this information for business development to find out where they should go. This is how 90 percent of the fed contracting game is won or lost— [with] knowledge [about] where to find that data."[55]

Distributed Spectrum's NSIN mentor recommended that the team apply to an NSIN program called Vector, which is intended to help NSIN alumni teams form a company and develop dual-use technologies. Distributed Spectrum was selected to participate in March 2021. Through Vector, an eight-week training program that culminates in a pitch event, teams learn business fundamentals and the basics of government contracting.[56] During the eight weeks, NSIN provides computer-based training covering business formation, market analysis, sales, fundraising, government contracting, and pitch preparation through an NSIN subcontractor, Dcode.[57] Distributed Spectrum participated in the Vector program and competed in the program's final pitch competition alongside eight other teams.

Distributed Spectrum focused its Vector pitch on the new general-purpose spectrum monitoring platform. As part of the accelerator, Distributed Spectrum explored adapting its technology to a variety of DoD use cases, including RF transmitter intrusion detection in secure environments. During Vector, the team interacted with a prospective end user, 1st Calvary Division at Picatinny Arsenal, and did a site visit to better understand how their Distributed Spectrum product might be used in a secure compartmented information facility (SCIF). A panel of judges from DoD and industry evaluated the Vector teams' technology solutions in a final pitch event. Distributed Spectrum won its Vector cohort and was awarded a $25,000 contract to support further development of the technology.

Over the course of the six-month Vector contract, Distributed Spectrum further developed its technology and received more guidance about interacting with government stakeholders and identifying potential clients. Following the two NSIN competition victories, Distributed Spectrum was introduced to an STTR phase I awardee, Teleqo Technical Solutions, through a contract at a venture capital business. Teleqo asked Distributed Spectrum to be a subcontractor on an STTR phase II award, which had been successfully funded in August 2020. Distributed Spectrum received a $200,000 subcontract to develop a wideband signal detection and classification system to pair with Teleqo's lidar scanning technology to map transmitters in urban environments.

In December 2021, Distributed Spectrum learned that it had been awarded the NSF SBIR phase I, more than a year after the team had initially applied. The company received $235,000 in February 2022 to develop a dense network of low-cost spectrum monitoring devices to

[55] Interview 30, Interviewee 40, September 29, 2021.

[56] National Security Innovation Network, "NSIN-Program Alumni Received $25,000 for DoD Development," June 3, 2021d.

[57] NSIN, 2021d; Carten Cordell, "Dcode Inks 5-Year Contract with Air Force Innovation Hub," *Washington Business Journal*, September 3, 2021.

distribute around Harvard's campus in Cambridge, Massachusetts. With this network, Distributed Spectrum can run additional experiments to test its system's signal detection and localization capabilities.

What Is the Potential Military Application and Market?

The system can detect intrusions in RF environments in both indoor and outdoor fixed areas of interest, such as SCIFs. The system can also be integrated with other mobile platforms such as vehicles to provide real-time monitoring and RF threat detection.[58]

Status: Are They Selling to DoD?

Distributed Spectrum has received a total of $250,000 in DoD contracts, including two $25,000 contracts through NSIN programs and roughly $200,000 as a subcontractor on the Department of the Air Force STTR phase II. The company received an additional $235,000 from the NSF SBIR phase I. Distributed Spectrum technology is currently at TRL 5. The company has built a prototype of its technology and validated it in a controlled setting.

Commercial Marketplace Story

What Is the Potential Commercial Application and Market?

Originally, Distributed Spectrum was going to tailor its technology to meet the needs of industries in dense RF environments. The intent was to help industries detect, identify, and locate interfering signals in the local area that conflict with the industry RF signals, such as in factory environments, warehouses, airports, and other industrial environments that rely on RF communication.

Status: Are They Selling on the Commercial Market?

The commercial market requires some level of technology maturation, and Distributed Spectrum has not yet formally worked with commercial entities. The company will use its contract with the NSF to conduct the basic R&D necessary for a commercial product, and it plans to apply for an NSF SBIR phase II to further development after the phase I work is complete.

Challenges

What Challenges Did They Face in Engaging with and Selling to DoD?

Distributed Spectrum noted that high barriers to entry to engage the federal government and the DoD pose a significant challenge, particularly for small start-ups. For example, many federal government organizations require an extensive proposal, which is prohibitively time-consuming for a start-up with limited resources, or they often require specific language and background knowledge to "read between the lines" of the solicitation. Distributed Spectrum

---

[58] Distributed Spectrum LLC, undated b.

further noted that simpler application processes and requirements will permit more start-ups to engage with the federal government and DoD.[59] Distributed Spectrum also observed that some opportunities appear to be framed with a specific company in mind to fill the DoD need, which makes it difficult for a new entrant to compete fairly. Finally, Distributed Spectrum stated that it can be difficult to identify and engage with prospective end users to understand their needs.

How Did They Address These Challenges?

Before engaging the innovation ecosystem, Distributed Spectrum was seeking to

- identify DoD opportunities with low barrier to entry (minimal resources required)
- identify customer identification and use cases (commercial and DoD)
- understand DoD-isms (culture and nomenclature issues)
- understand DoD strategic direction
- understand how to navigate DoD bureaucracy
- have direct interaction with end users to inform technology development
- develop business and marketing acumen (e.g., pitch skills, writing proposals, overall business acumen).

NSIN helped Distributed Spectrum meet some of these needs. NSIN provided a low-barrier-to-entry way to compete for limited funding. Perhaps more critically, NSIN provided introductions and access to DoD end users and training and navigation support to help the Distributed Spectrum team understand DoD acquisition processes. Through their NSIN mentor and NSIN-facilitated engagement with DoD end users, Distributed Spectrum came to better understand military nomenclature, buzzwords, strategy, and systems, which has helped the company learn to "speak DoD" and avoid cultural mistakes that could impede effective communication with end users. In addition, engagement with end users has helped the Distributed Spectrum team tailor their product descriptions and product development to the department. For example, through this engagement, the team learned that DoD is moving toward a Joint All-Domain Command and Control concept, which requires systems to be able to interface with different types of systems. Distributed Spectrum's RF monitoring system is modular and compatible with other systems, and after engaging with end users through Vector, the team understood that in framing their concept in DoD language—using the JADC2 phrase—they can help DoD customers to understand how Distributed Spectrum technology aligns with the strategic direction of the department. Distributed Spectrum also gained a better understanding of general business practices and the commercial industry.

---

[59] Interview 20, September 15, 2021.

Distributed Spectrum notes that partnering with the NSIN mentor has been instrumental in the team's progress over the last year, helping them learn business development and better understand the DoD ecosystem and negotiation tactics.[60] According to the team,

> [engagement with] NSIN has been critical to our success as an early-stage startup. Through the Hacks and Vector programs, we landed our first government contract, made valuable connections throughout the DoD, and gained advisors who helped us tremendously to navigate the many challenges of beginning to work with the government.[61]

Even with robust support and training, the pathway to development of the technology and adoption by DoD is unclear. DoD customers have been interested in the Distributed Spectrum technology, but the company has not yet identified a customer willing to support the continued maturation and development of the technology.

The team still needs opportunities to validate their technology, including testing and proof-of-concept opportunities, and a pathway to see their technology transitioned to DoD for use.

What Is Next for the Company?

In the summer of 2022, Distributed Spectrum was focused on fulfilling the AF phase II STTR subcontract and NSF SBIR phase I contract. Distributed Spectrum's founders were graduating from Harvard in May 2022. After graduation, the founders will move company operations from Cambridge to New York City and continue to work on the business full-time.

One of Distributed Spectrum's primary objectives is to secure its own DoD SBIR or STTR contract from which it can begin to build relationships with stakeholders that can help the company test its technology in an operational environment. Additionally, Distributed Spectrum will continue R&D activities to help prepare its product for customer demonstrations and other engagements. Finally, Distributed Spectrum plans to participate in demonstration events such as the Naval Postgraduate School's Joint Interagency Field Experimentation Program (JIFX) and hopes to gain entry to accelerator programs such as NSIN Propel and MassChallenge that will help the company further develop as a business.[62]

---

[60] Interview 20, September 15, 2021.

[61] NSIN, 2021d.

[62] NSIN information to authors, April 4, 2022; Naval Postgraduate School, "NPS Field Experimentation," webpage, undated; MassChallenge, homepage, undated.

Small Business ("Company X")—Small Unmanned Surface Vessel

Technology Story

What Is the Technology?

The company (which asked for anonymity) is an R&D-focused organization specializing in a range of technologies and design services, including advanced hydrofoil hulls; unmanned marine and land vehicles; and motion-compensated land, air, and marine vehicle launch and recovery systems. Two technologies that Company X developed, a small surface vessel and an elevated mast, were successfully demonstrated by the U.S. Navy in 2019. The vessel is a multi-mission platform based on a commercial personal watercraft design that is available in both manned and unmanned configurations and was designed for durability in coastal waters characterized by breaking surf and high waves. According to a specification sheet obtained by the authors, the elevated mast is a large, tethered parafoil kite "designed to deliver dramatic line-of-sight expansion to maritime assets." Both were created specifically for military applications—however, both also have various commercial applications.

What Is the Company, and Who Are the Founders?

Headquartered in the mid-Atlantic, Company X was started in the 1980s by a former Naval Surface Warfare Center civilian interested in developing and fielding maritime technology faster than usually observed in government processes. The founder initially pursued consulting work and engineering services, eventually forming today's company. With the support of the SBIR program, the company was introduced to the hardware world and gradually moved into design and manufacturing, which is its main focus today. Currently, the company seeks to "transition technologies to production" and provide capabilities and support services for a wide range of systems on sea, land, and air, from concept through production and field support. The company aims to be about 50 percent commercially focused and 50 percent DoD focused.[63] The actual percentage fluctuates, with some years ranging from 60 to 70 percent commercial, but most years the company has been around 80 to 90 percent DoD focused.[64] The company has over 20 years of experience with unmanned surface vessel (USV) work. It developed its first unmanned marine vessel in 1996 with the aid of DoD S&T funding.

What Does the Technology Do?

The company has developed and designed a wide variety of products, from crew transfer vessels to life raft pods. One of the products, a small maritime vessel, is a multi-mission modified watercraft platform. The original platform was a foreign design to rescue people caught in the

---

[63] Interview 49, interviewee 66, January 13, 2022.

[64] Interview 49, interviewee 67, January 13, 2022.

surf.[65] Former SOF operators discovered the design and planned to sell it to the Air Force but did not have the finances or infrastructure to continue development, so design rights were eventually purchased by the company,[66] which then built 24 boats for the Air Force to drop out of C-130s as sea rescue crafts and aimed to make a diesel boat made out of aluminum.[67] According to a specification sheet provided by the company, the vessel is designed for durability and high surf and wave conditions and comes in a wide range of variations, including "unmanned/manned, gas/electric/diesel, and 3.5/4.3 meter" boats. There are also variants for specific uses, such as search and rescue (SAR). When manned, the vessel can be used to rescue those stranded in deep water. The company has also developed unmanned configurations for both military and commercial applications. According to the specification sheet, the unmanned variant is remote controlled and used commercially for hydrographical survey and ocean data collection, optimized for extreme weather and ocean conditions. The military unmanned version has the same capabilities and is rated as a Class 1 USV platform designed for defense and security missions.

The vessel is used in tandem with an elevated mast, also developed internally. The mast is a tethered airborne platform developed to deliver greater and more efficient line-of-sight expansion to maritime vehicles and radar and radio relay communications under various wind and weather conditions. As a parafoil-based kite, it can be attached to the small USV (sUSV). It can carry payloads from the host vessel to 500 feet while sustaining bidirectional communications with the ship, powering up to 50-pound payloads, and performing intelligence, surveillance, and reconnaissance (ISR) functions. The larger elevated mast is capable of raising 100-pound payloads to 1,000 feet.

DoD Story
How Did They Get Introduced to DoD?
As previously noted, the company was founded by a former Navy civilian who had prior experience and familiarity with DoD. The SBIR program was instrumental in aiding the company's shift from consulting and engineering work into hardware services. The company recalls that learning how to do business with both DoD and the commercial market was "strictly trial and error."[68] According to interview comments from the company, in 2004, a former tech director of one of the Navy's warfare centers approached them about the possibility of developing high-speed unmanned boats for the Navy, and through their collaboration, they designed and built two unmanned boats. The company tried to pursue fleet transition with the Naval laboratory system, but it became clear that the boats were not going to come out of the laboratory system and were instead to become test vessels because

---

[65] Interview 49, interviewee 66, January. 13, 2022.

[66] Interview 49, interviewee 67, January 13, 2022.

[67] Interview 49, interviewee 66, January 13, 2022.

[68] Interview 49, interviewee 66, January 13, 2022.

there was no demand pull from the fleet or documented requirement. At the time, the company also noted that there were no Navy program offices set up to fund unmanned surface vehicles. The company made extensive use of independent research and development (IRAD) funding to continue design and development work during periods when DoD funding was not forthcoming.[69]

What Is the Potential DoD Market Problem Set?

The sUSV with elevated mast addresses a wide market for the DoD. The vessel falls into the category of an USV while the mast is a tethered UAV. Both address a plethora of DoD problem sets, including SAR operations, hard-to-reach terrain exploration, communications and data relay during inclement weather conditions, and ISR functions.

Interaction with Innovation Ecosystem

The company has frequently interacted with the DoD innovation ecosystem. It contracted with a defense innovation organization as part of a joint program between DoD and the U.S. Navy. The program successfully tested the elevated mast in 2015, with over 20 flights launched under various wind and weather conditions on different platforms. The mast also underwent a series of demonstrations on board both manned and unmanned Navy vessels. It took a total of ten months to go from the demonstration concept to the actual at-sea demonstration.

In 2017, the company received a Navy SBIR award to demonstrate a configuration suitable for extending the communication and control range of unmanned systems beyond line of sight. Also in 2017, the company worked with another DIO to develop a prototype of its technology concept, continue R&D, and conduct a demonstration. The company had proposed to develop and demonstrate (both deploying and retrieving) an autonomous parafoil airborne system with a ten-pound payload, with its small vessel allowing for longer-range communications and increased sensor connectivity than line-of-sight constrained systems and with the proposed system capable of autonomous launch, coordination, operations, and recovery. The integration of the two systems would increase the capability and number of USVs for the sea services. The DIO approved the proposal and invested over $2 million to fund a prototype, support R&D to mature the technology, and then fund a demonstration. The DoD funding was applied to the existing SBIR contract as the mechanism to get the funds to the company.

The company also participated in several live exercises with versions of the vessel and elevated mast. These exercises allowed the company to demonstrate both the feasibility and utility of its technology and systems to a range of potential DoD customers. Funding for participation in each exercise came from the DoD sponsor of the exercise.

The company was able to transition the sUSV/mast system to the Navy. Plans included producing 48 systems in support of an existing program of record. Because of funding priorities and delays in the larger program of record, the Navy resource sponsor shifted that

---

[69] Interview 49, interviewee 66, January 13, 2022.

requirement from the initial operational capability (IOC) date to a future full operational capability (FOC) date.[70] So, while the sUSV/mast system is planned to be part of a program of record, and the Navy is not looking at alternatives that might also meet the beyond line-of-sight communication requirement, there is currently no funding to support further development and production of the system.[71]

Commercial Marketplace Story

What Is the Potential Commercial Market/Problem Set?

The sUSV/mast system is dual use and has commercial applications, specifically in the maritime industry. For example, the small vessel has a commercial rescue boat version dedicated for search and rescue. The larger elevated mast can also be used for increased sensory detection for marine wildlife. The company has targeted the commercial offshore wind industry as a potential market, specifically using a sensor carried by the mast that increases the range at which marine wildlife, and endangered whales in particular, can be detected and therefore avoided by the boats carrying operations and maintenance personnel to the wind platforms. To demonstrate this capability, the company found an opportunity for an SBIR award outside DoD.

Challenges

What Challenges Did They Face in Engaging/Selling to DoD?

The company noted two main sets of challenges:

1. Time to get on contract. With the notable exception of the last set of Navy SBIR funding, other contracts could take as much as 12 months from the time an award was indicated to the actual contract award.
2. Navigating the DoD labyrinth of organizations responsible for technology development, demonstrations and exercises, contracting, requirements, technology transition, and program management. Even with insider knowledge of DoD, the company acknowledged that there are many relevant potential customers with varying interests that need to be understood and addressed in order to obtain opportunities for funding to support further technology development and maturation.

How Did They Address These Challenges?

To address challenges in delay, the company stated that it simply "overbooked . . . in terms of writing proposals because you know this is going to happen on a regular basis."[72] The

---

[70] Interview 50, interviewee 69, January 13, 2022.

[71] Interview 49, interviewee 66, January 13, 2022.

[72] Interview 49, interviewee 66, January 13, 2022

company also used IRAD funds to cover gaps in DoD funding opportunities and the time it took DoD to make an expected contract award. This is particularly relevant because the company chose not to lay off people during times of low revenue.

The company did not do anything in particular to address the navigation and multiple possible customer organizations. Its prior knowledge and current experience working for and with DoD enabled the company to navigate among DoD offices.

What Is Next for Them?

As of today, the sUSV/mast system is essentially a mature technology at approximately TRL 7 that requires some additional R&D work, as well as integration work, to be a fully capable system.[73] The expectation was that this would be accomplished as part of the Navy program of record.

Company X is also turning its focus toward the commercial market—and continuing toward its interest of going at least 50 percent commercial. The company has targeted the offshore wind market and intends to commercially introduce its maritime technologies. It has approached a large offshore wind developer with a proposal to partner and build on the phase I non-DoD SBIR work in using the elevated mast to detect endangered whales.[74]

---

[73] Interview 49, interviewee 66, January 13, 2022.

[74] Interview 49, interviewee 66, January 13, 2022.

Vita Inclinata

Technology Story

What Is the Technology?

Vita Inclinata's technology is described as "the world's first autonomous stability system that lets you place a load, hoist, or litter exactly where you want it."[75] All of the company's products are designed with its proprietary Load Stability System (LSS), which consists of sensors and software that run computations 100 times per second to tell a lift system's thrusters when to stop spin and swing, or how to orient a load within seconds. The LSS was designed with input from pilots, medevac crews, and crane operators, with support from defense innovation organizations, primarily AFWERX.

Vita has developed four products: (1) the load pilot, (2) the load navigator, (3) the litter attachment, and (4) the sling load. Vita is in the process of developing an additional two products: (5) the hoist attachment, and (6) the precision rapid aerial employment system (PRAES). All products appear to have dual-use potential, are battery-powered, work under most helicopters and cranes, are maintenance-free, and Army electromagnetic interference (EMI)-certified to not interfere with aircraft electronics.

The Vita load pilot is a remotely operated suspended load system capable of lifting up to 12 tons in the most challenging environments.[76] The Vita load navigator is a heavy-duty stabilization system for cranes capable of carrying up to 40 tons of weight, depending on configuration and whether it is mounted to telescoping spreader bars.[77] The Vita rescue system litter attachment is a lightweight, quick-attach unit that is compatible with a number of rescue kits. It is meant to allow helicopter crews greater speed, safety, and control on hoisting operations. The Vita sling load is built for long line rotary-wing operations and can be configured to stabilize sling load swing, or spin and rotation. Depending on configuration, it can carry up to 25,000 pounds.[78] The hoist attachment and PRAES are both under development. The PRAES is described as a dual-use solution for commercial and military operators to conduct fixed-wing extractions without putting people or loads at risk.[79]

Vita describes its technology as the first platform-agnostic, autonomous, suspended-load stability system with both earth and space applications. LSS provides reengineered drone technology, a thrust-based solution to afford more control of cable-slung loads. This decouples

---

[75] The technical descriptions of the technology are taken primarily from Vita Inclinata, "Vita's Systems: LSS Technology Products," webpage, undated d.

[76] Vita Inclinata, "Vita's Systems: LSS Technology Products: Load Pilot," webpage, undated h.

[77] Vita Inclinata, "Vita's Systems: LSS Technology Products: Load Navigator," webpage, undated g.

[78] Vita Inclinata, "Vita's Systems: LSS Technology Products: Litter Attachment," webpage, undated f.

[79] Vita Inclinata, "Vita's Systems: LSS Technology Products: Hoist Rescue," webpage, undated e; Vita Inclinata, "Vita's Systems: LSS Technology Products: Precision Rapid Aerial Employment System (PRAES)," webpage, undated i.

motion between a load and an aircraft. According to a description, "a series of fans automatically provide counter thrust in any direction to a swinging load independent of the flying crew. A center module containing inertial measurement units (IMUs), gyros, and other sensors directly communicates to the fans, providing full control of any swinging load. Furthermore, the device can be applied to any cable-based suspended load system including crane operations, sling load operations, drone delivery systems, firefighting operations, and deep space operations."[80]

What Is the Company, and Who Are the Founders?

Caleb Carr was a volunteer tech for Multnomah County Search and Rescue in Oregon during his youth. In 2009, during a routine night search-and-rescue training mission, Carr's friend and guide went into cardiac arrest. A rescue helicopter arrived on scene, but gusty winds and mountainous terrain prevented Carr's friend from being saved, despite multiple attempts to extract him. Similar situations lead to thousands of lives being lost every year.

Carr attended the University of Colorado at Denver, graduating with degrees in neuroscience and public policy. While there, he met Derek Sikora, who would later become Vita's cofounder and chief technology officer (CTO). Though Carr originally intended to become a medical doctor to save lives, he was persuaded by a professor to try to fix the problems that led to his friend's death. This would drive Vita's mission to create a solution that senses the environment and immediately adjusts to stabilize load swing and spin for rescue crews and crane operators.[81] Vita was founded in 2012 and, beginning in 2018, strove to perfect the technology by working with various defense incubators and going through competitions.[82]

Carr later earned a law degree from the Mitchell Hamline School of Law, focusing on corporate law and litigation. He has litigated over 300 administrative law cases to date. He is currently a student at Penn State University, completing a master's degree in business administration.[83]

How Was the Technology Developed?

Vita's website says the LSS was designed with input from pilots, medevac crews, and crane operators.[84] Employees have backgrounds in the aerospace industry and have worked at NASA, Microsoft, and Uber. After research, the team found that rescue baskets suspended from helicopters can swing by as much as 25 degrees from vertical, depending on the weather and the helicopter itself. As of 2019, the team had worked with the Air Force, Army, and Navy

---

[80] Business Journals Content Studio, "Vita Inclinata Technologies: winner of the Entrepreneurial Opportunity Contest," *Denver Business Journal*, September 13, 2019.

[81] Vita Inclinata, "About: Building Technology That Brings People Home Save Every Time," webpage, undated a.

[82] LinkedIn, Vita Inclinata Technologies, undated c; Interview 25, Interviewee 32, September 21, 2021.

[83] Vita Inclinata, "Caleb Carr," webpage, undated b.

[84] Vita Inclinata, undated d.

to develop its autonomous system. By that point, Vita had built an initial prototype that was tested on three different helicopters for a total of 12 successful flight hours and had undergone significant lab testing.[85]

DoD Story

How Did They Get Introduced to DoD?

Vita's exposure to the defense innovation ecosystem came about by happenstance. Carr's wife was in the USAF and had some professional connections to AFWERX. Early team members who had prior service in or other connections to the military initiated contact with some parts of the wider defense community, including Air Force, Army, and National Guard units. Vita brought individuals familiar with DoD customers and the defense acquisition process in as employees or advisers to expand Vita's customer base and help transition to a program of record through their insight and expertise.[86]

Interaction with Innovation Ecosystem

All of Vita's products have been incubated through AFWERX, though Vita has also worked with Army xTechSearch and a few other smaller DIOs. Vita won first place and $250,000 at xTechSearch, a program that selects small businesses to provide proof-of-concept of dual-use technologies that can apply to Army missions and seeks to integrate these businesses into the Army's S&T ecosystem.[87] Vita was also a part of the fall 2018 NSIN (then named MD5) cohort after responding to an application on manned-unmanned teaming that awarded a phase I SBIR contract. Vita was one of the few ventures that transitioned to phase II; in fact, Vita received three phase II contracts through 2019 and 2020.[88] These small infusions helped to develop early iterations of the technology and product.

Vita also interacted with Air Force Accelerator Powered by TechStars, whose managing director, Warren Katz, currently sits on Vita's board of directors.[89] While this support is credited with helping to set the company up for success, Vita encountered a number of challenges (discussed further below). It has gone through five incubators, including AeroInnovate, which Vita later purchased in 2019.[90]

---

[85] Business Journals Content Studio, 2019.

[86] Interview 25, Interviewee 32, September 21, 2021.

[87] Jonathan "jmill" Miller, "Bringing Heroes Home, Featuring Caleb Carr of Vita Inclinata," *Tough Tech Today with Meyen and Miller*, November 3, 2020; Michael Howard, "xTechSearch 4 Grand Prize Winner's Technology Offers Innovation in Helicopter Rescue Operations," October 22, 2020; xTech, "xTechSearch 4," webpage, undated c.

[88] Interview 32, Interviewee 43, October 12, 2021, and citing an AFWERX portfolio spreadsheet, contract numbers FA86491999009, FA86501999335, and FA86492099097.

[89] Vita Inclinata, "Our Team," webpage, undated c.

[90] Miller, 2020.

Vita was awarded its first DoD contract in November 2018 as a SBIR phase I from AFWERX for its Automated Helicopter Suspended Load Stability System for Hoist Rescue Operations. This first SBIR phase I award and participation in the Air Force Accelerator led to three SBIR phase II contracts.[91] Vita won a 2020 SBIRS-STTR Tibbets Award, which recognizes "those companies, organizations, and individuals that exemplify the very best in SBIR/STTR achievements."[92]

In all, MD5 provided first $125,000 in phase I, then $1.5 million in phase II funding. AFWERX provided $75,000 in phase I, then $1.5 million for phase II. This funding helped develop the technology, but the next steps were driven by the Vita team.[93]

What Is the Potential Military Application and Market?

Vita's current defense market is primarily Army and Air Force search-and-rescue and cargo operation support. This market includes Army and Air National Guard units. Navy, Marine Corps, and Coast Guard units performing similar missions are potential future customers. To communicate a demand signal for its products, the Vita team adapted an AFWERX memorandum of understanding (MOU) template and got over 100 signatures from personnel (or units) across the DoD to signal demand for Vita's product. Through this set of engagements, Vita also learned about the DoD acquisition process.[94] Vita says at least 20 military units have signed letters of support for its product, signaling demand for this capability to DoD and service headquarters.[95]

Status: Are They Selling to DoD?

In the last few years, Vita has built a robust government relations division to facilitate interactions with Congress. Working with Senate and House Armed Services Committee authorizers, specifically Sen. Jim Inhofe (R-Okla.) and Rep. Kendra Horn (D-Okla.), Vita was able to persuade Congress to authorize and appropriate $5.5 million for the Air Force to purchase its technology. Congress appropriated a further $5.37 million in FY 2022. Vita officially executed a contract in October 2021 with the Defense Logistics Agency's Tailored Logistics Support (DLA TLS) program and expects to become a contractor for a program of record within the next two to three fiscal years.[96]

---

[91] Miller, 2020.

[92] U.S. Small Business Administration, SBIR/STTR America's Seed Fund, "Tibbets and Hall of Fame," webpage, undated c.

[93] Interview 32, Interviewee 43, October 12, 2021.

[94] Interview 32, Interviewee 43, October 12, 2021.

[95] Interview 25, Interviewee 31, September 21, 2021.

[96] Interview 32, Interviewee 43, October 12, 2021.

Commercial Marketplace Story

What Is the Potential Commercial Application and Market?

Vita's technology is dual-use and can apply to industries lifting unstable loads. Vita identified prospective commercial applications in construction, oil, and firefighting and continues to look for other potential uses. Carr has said that "cranes were a massive market opportunity that we didn't even think about. I never thought about construction. Now I really care about construction. And the reality is that this industry had the exact same problem of being able to stabilize loads and people being injured. We would have had no idea if it weren't for looking at the fringe opportunities."[97]

While working with DIOs, Vita sought private capital. The company notes that it only went through one accelerator, TechStars (which is affiliated with AFWERX), but raised $17 million in private venture funding.

Status: Are They Selling on the Commercial Market?

Vita is selling to the construction industry, particularly crane operations. Vita is also working in the wildland fire equipment market.[98] Vita has work within the civilian search-and-rescue sector. The company also noted that it is doing international work with the Israeli and Indonesian air forces, as well as the Japanese police and self-defense force. Vita noted that because the SBIR applications from NSIN and AFWERX required dual-use technology, it forced the company to think about a broader set of uses for Vita's technology.[99]

Challenges

What Challenges Did They Face in Engaging with and Selling to DoD?

Vita encountered challenges with business development and finding customers within the DoD. Vita also struggled with the acquisition and contracting process. In general, the Vita team indicates there is a lack of transparency with how the DoD operates and that breaking through requires some insider knowledge. Although AFWERX provided initial connections and contacts, its approach was generally hands off. According to Vita, AFWERX can improve by connecting companies from accelerators to potential DoD customers and units.[100] As one interviewee said, "Luck and a few connections made a big difference."[101]

The interviewee suggested that more mentorship could be helpful. The process gets harder once a company has to get contracts, and as the process gets harder, "they [DIOs] get less

---

[97] Miller, 2020.

[98] Miller, 2020.

[99] Interview 32, Interviewee 43, October 12, 2021.

[100] Interview 25, Interviewee 31, September 21, 2021.

[101] Interview 25, Interviewee 32, September 21, 2021.

involved. They don't tell us we can go talk to units. I think they can do much better. It is not handholding, but DoD is a big scary beast. They're like, 'Go figure it out.'"[102]

In a 2020 podcast interview, Carr gave a candid assessment of the strengths and weaknesses of incubators, stating that they are great at teaching and training entrepreneurs to be business professionals. They play an important role early on by making connections, providing funding, and helping companies find their markets, but they provide less value after that point in a company's development.[103] After the program is complete, DIOs have little follow-through, handoff to the next stage, or sustained effort to help early-stage ventures find and engage customers.

Vita also noted that small businesses interacting with DoD believe that they have to interact with units, general officers, or program managers to get traction, but "that's not the case."[104] Vita suggests that businesses have to understand or generate military requirements for their solution or product, then try to drive money toward that.

How Did They Address These Challenges?

Vita found a "Blue Book of Aviation" and used it to find potential customers within the Army that might need Vita's technology. Vita started cold-calling DoD units and pitching them its products. Vita also hired individuals with defense acquisition backgrounds to help navigate the acquisition process. Vita believes they are on the precipice of cracking the shell and breaking into major acquisition programs. Vita also cold-called and introduced the company to several members of Congress to advocate for hoist stability technology and get a line item in the DoD budget.

Vita also successfully circumvented, for lack of a better word, the normal acquisitions process. Vita approached Congress and was able to get $10.87 million appropriated for DoD to purchase its technology. "What we also learned is that when you get money from Congress, it rapidly accelerates the process. We went through testing processes in nine months. That is unheard of for something on aircraft. We're bringing money. You [DoD] don't have to think about it, we'll do that; you [DoD] just need to support the capability."[105]

What Is Next for the Company?

Vita expects to be on a POR in FY 2024 or FY 2025 for its load stability system. The company is also working to get traction with the Army's PEO Soldier. Its product is on the unfunded requirements list for the FY 2024 program objective memorandum (POM), and the company has been working with Army PEO Aviation to ensure it is well situated for FY 2025.

---

[102] Interview 25, Interviewee 32, September 21, 2021.

[103] Miller, 2020.

[104] Interview 32, Interviewee 43, October 12, 2021.

[105] Interview 32, Interviewee 43, October 12, 2021.

Lessons for Defense Innovation Organizations

Based on their experiences with DIOs over time, the Vita team observed that there has been little improvement in the defense innovation ecosystem. Innovation organizations focus on getting small businesses into the defense ecosystem, but once these small businesses are in the process, innovation organization support dissipates. "We won xTech; hypothetically, this is the golden child of Army procurement. But short of the $250,000 prize money, nothing else came of it."[106]

Vita advises that DIOs help small businesses create an operational need statement (ONS) and related documents to drive conversations with service acquisition authorities. "The more you submit, the more you drive the conversation" and the more the system will begin looking at service needs.[107] An interviewee suggested that the PM medevac did not know there was an operational need, even though Vita generated at least 30 MOU documents signed by Air Force and Army units that did not commit the unit to procurement but did signal a demand for the product. At that level of support, the PM would have to have a conversation with Vita about the capability its products provide.[108]

---

[106] Interview 32, Interviewee 43, October 12, 2021.

[107] Interview 32, Interviewee 43, October 12, 2021.

[108] Interview 32, Interviewee 43, October 12, 2021.

Xona Space Systems

Technology Story

What Is the Technology?

Xona Space Systems seeks to "provide global navigation services with the accuracy, security, and resilience needed to operate systems safely in any environment."[109] Xona is building Pulsar, "a precision position, navigation and timing (PNT) service using small satellites in low earth orbit (LEO)."[110] LEO satellites operate approximately 500 to 1,000 km in altitude from earth, unlike modern-day legacy Global Navigation Satellite Systems (GNSS), including GPS in the United States and Galileo in the European Union (EU), which operate in medium earth orbit (MEO), approximately 20,000 km from earth.

This technology addresses the gap between the growing PNT demands from civil and government end users and the capabilities that traditional GNSS offer. GNSS is made up of constellations of navigation satellites that provide "signals from space that transmit positioning and timing data to receivers, which then use this data to determine time and location."[111] For example, GPS, whose primary focus is military applications, relies on 30 operational MEO satellites operating at approximately 20,200 km from earth.[112] GPS is the world's most utilized satellite navigation system, with land, air, sea, and space applications.[113]

GNSS systems have become the backbone of nearly every aspect of the modern connected world. GNSS has generated trillions of dollars in economic benefits since it was made available for commercial use in the 1980s.[114] Xona notes that in the United States alone, "over $300 billion of economic impact per year" relies on GNSS.[115] However, because the signals are so far away, they are weak and easy targets for interference. GNSS is vulnerable to spoofing, jamming, and cyberattacks.[116] GPS signals that are available to civilians are

---

[109] Xona Space Systems, "Background," webpage, undated b.

[110] Xona Space Systems, "Xona Pulsar," webpage, undated c.

[111] European Union Agency for the Space Programme, "What Is GNSS?" webpage, undated.

[112] There were a total of 30 operational satellites in the GPS constellation. This number varies; the system can operate with as few as 24 satellites. GPS.gov, "Space Segment," webpage, undated c.

[113] GPS.gov, "GPS Overview," webpage, undated b.

[114] Kathleen McTigue, "Economic Benefits of the Global Positioning System to the U.S. Private Sector Study," NIST, October 2, 2019.

[115] Xona Space Systems, "Xona Space Systems Raises $1M Pre-Seed Round from 1517, Seraphim Capital, Trucks Venture Capital and Stellar Solutions," press release, May 14, 2020.

[116] Kate Murphy, "America Has a GPS Problem," *New York Times*, January 23, 2021; Guy Buesnel and Mark Holbrow, "GNSS Threats, Attacks and Simulations," SPIRENT, June 2017; Yoav Zangvil, "Research on GPS Resiliency and Spoofing Mitigation Techniques Across Applications," presentation, undated. Xona also notes that "the only [GNSS] signals available to the public are unprotected, imprecise, and susceptible to disruption from easily obtainable jammers and spoofers" (Xona Space Systems, 2020b).

not encrypted.[117] Furthermore, GNSS is vulnerable to unintentional disruptions, including atmospheric interference.[118] With emerging applications such as self-driving vehicles, consumers are also increasingly demanding enhanced performance that GNSS was not originally designed for and may not deliver.[119]

Leveraging recent advances in satellite technology and their patent-pending system architecture, Xona Pulsar provides users with a resilient and global alternative navigation (AltNav) solution that is fully independent from GNSS.[120] By operating 20 times closer to the earth than GPS satellites, Pulsar offers "10 times more accuracy than current Global Navigation Satellite Systems through encrypted signals and 100 times better interference mitigation."[121] Pulsar's encrypted and jam-resistant signals with rapid convergence times from LEO offer stronger signal power than traditional GPS.[122] While more satellites are needed to establish a functional LEO PNT architecture, their closer proximity to end users and faster satellite motion across the sky enable significant advances over traditional GNSS.[123]

The National Defense Authorization Act for FY 2021 signaled Congress's growing interest in better PNT capabilities for government and military as well as civilian uses.[124] For both government and commercial applications, Xona Pulsar can assist with resilient positioning for any end user over land, sea, air, and space.[125] Xona's product has wide applicability commercially, including, but not limited to, critical infrastructure (power utilities, telecommunications, financial), intelligent transport systems such as self-driving vehicles and drones, robotics, smartphones, and internet of things (IOT) devices.[126]

---

[117] Jason Rainbow, "Xona Space Systems Fully Funds GPS-Alternative Demo Mission," *SpaceNews*, September 22, 2021.

[118] Murphy, 2021.

[119] The U.S. government commits to broadcasting a GPS signal with a user range error of roughly 2 meters (m), though it reports that 95 percent of the time, actual performance is closer to 0.6 m. User accuracy depends on other factors, however, including atmospheric conditions and receiver design features (GPS.gov, "GPS Accuracy," webpage, undated a). "Urban canyons, tree-lined streets, tunnels, and underpasses can all block GNSS satellite signals long enough to interrupt the positioning" (Dan Dempsey, "11 Myths About GPS for Autonomous Vehicles," *Electronic Design*, May 8, 2019).

[120] Newlab, "NSIN Propel Demo Day," Bryan Chan, Xona Space Systems cofounder presentation, video, Youtube, September 27, 2021.

[121] Inside GNSS, "LEO PNT Service Provider, Equipment Manufacturer Team for Pulsar PNT," November 19, 2021.

[122] Xona Space Systems, 2020b.

[123] Interview 39, interviewee 50, October 27, 2021; Xona Space Systems information to authors, April 1, 2022.

[124] Dana Goward, "2021 Defense Act Signals Turning Point for Congress and PNT," *GPS World*, December 29, 2020.

[125] Newlab, 2021.

[126] Newlab, 2021.

What Is the Company, and Who Are the Founders?

Xona was founded in 2019, in San Mateo, California, by Stanford Aerospace Engineering PhD and master's graduates.[127] Dr. Tyler Reid, one of the cofounders, worked in the Stanford GPS lab during his doctoral studies. Part of his dissertation work, funded by DARPA, focused on using LEO satellites to augment GPS.[128] After Reid graduated from Stanford, he worked at Ford Motor Company's autonomous vehicles division as a navigation expert, where he investigated how Ford could implement novel navigation sensors and technologies for cars to drive themselves. At Ford, Reid realized that GPS performance could not meet the rigorous demands of autonomous vehicles and that LEO satellite-based PNT might better meet their needs. Reid met with a group of Stanford graduates at night, trying to figure out if the idea to create a new service and company based on LEO technologies was technically and commercially feasible.[129] They created Xona Space Systems in April 2019.

The company's eight cofounders are graduates of Stanford with either a PhD or master's degree.[130] Currently, Xona has over 30 employees with prior work experience at companies including SpaceX, Ford Motor Company, Blue Origin, Booz Allen Hamilton, and other businesses.[131] The advisory board includes leadership from a retired Lockheed Martin executive, a Stanford University professor, a senior space systems architect from Stellar Solutions, a former executive from Beats by Dre, the founders of semiconductor maker u-blox, a former CEO of the U.S. Geospatial Intelligence Foundation (USGIF), and a retired U.S. Navy rear admiral.[132] Currently, Xona has set up subsidiaries in the United Kingdom and Canada, with plans to take on more employees in both countries.

How Was the Technology Developed?

Initially, Xona faced skepticism when its founders tried to explain how LEO satellites might be used for PNT services. In 2017, few researchers were exploring the idea of LEO satellite-based PNT services to augment GPS, and the idea was novel and not widely accepted as a technically feasible solution.[133] A Xona representative noted that this sentiment quickly changed in 2020, when the UK government was forced to reevaluate its approach to PNT.[134] After Brexit's effective date of January 2020, the United Kingdom was locked out of Galileo, the EU version of GPS, and the UK government recognized that it had to find an alternate

---

[127] Interview 39, Interviewee 50, October 27, 2021.

[128] Reid, Tyler Gerald Rene, *Orbital Diversity for Global Navigation Satellite Systems*, Palo Alto, Calif.: Stanford University, June 2017.

[129] Interview 39, Interviewee 50, October 27, 2021.

[130] Xona Space Systems, "About," webpage, undated a.

[131] Xona Space Systems, undated a.

[132] Xona Space Systems information to authors, April 1, 2022.

[133] Interview 39, Interviewee 50, October 27, 2021.

[134] Interview 39, Interviewee 50, October 27, 2021.

source of PNT.[135] The United Kingdom considered building its own PNT constellation but also explored commercial alternatives.[136] In July 2020, the UK government announced that it had bought nearly 50 percent of a company called OneWeb that was building LEO satellites for communication, with observers suggesting that the government intended to use OneWeb LEO satellites for PNT.[137] Xona noted that the UK government's decision "perked up some ears. All of a sudden, people started to look at who else was doing this, and how to implement it. Suddenly there was further interest in people talking to us."[138]

DoD Story
How Did They Get Introduced to DoD?
No information to report.

Interaction with Innovation Ecosystem
From 2019 to 2022, Xona received funding and support from several defense organizations, including DARPA, the NGA, the U.S. Navy SBIR program, NSIN, and the Naval Postgraduate School, as well as funding from the NSF. In mid-2020, Xona received a SBIR phase I award of $225,000 from the NSF to conduct R&D work on an "Atmospheric, Mapping and Satellite Integrity Monitoring (AMSIM) system to support a high-integrity, precise positioning service from a low earth orbit (LEO) satellite navigation constellation."[139] In early 2021, Xona received a SBIR phase I award of $140,000 from the U.S. Navy to conduct research on the feasibility of using Xona's LEO PNT service for hypersonic and LEO space vehicles.[140]

Xona connected with the NSIN through the NGA accelerator. After Xona had participated in the NGA accelerator, NSIN wrote Xona to invite them to apply for the NSIN Propel program. Xona applied, was accepted, and participated in NSIN's Propel program in June 2021. The program provided guidance for early-stage technology start-ups on developing traction with government agencies and explained detailed aspects of commercial-government relationships such as intellectual property rights and contract vehicles. One of the requirements of the NSIN Propel program acceptance was to test its technology in a real-life scenario.

What Is the Potential Military Application and Market?
See "What Is the Technology?" above.

---

[135] Jeff Foust, "Britain Charts a New Course for Satellite Navigation," *SpaceNews*, August 24, 2021.

[136] Resilient Navigation and Timing Foundation, "OneWeb PNT Possibilities," blog post, April 19, 2021.

[137] Inside GNSS, "UK Acquires OneWeb LEO Constellation, but Won't Work for SatNav—or Maybe It Will," July 14, 2020.

[138] Interview 39, Interviewee 50, October 27, 2021.

[139] Xona Space Systems, "Xona Space Systems Awarded Competitive Grant from the National Science Foundation," press release, December 21, 2020.

[140] Xona Space Systems information to authors, April 1, 2022.

Status: Are They Selling to DoD?

Xona Pulsar is currently at TRL 6 and has not yet sold product to DoD. DoD may be interested in applications of Xona's technology for high-performance alternative navigation systems with global availability that would function in GPS-denied environments.

Commercial Marketplace Story

What Is the Potential Commercial Application and Market?

Xona cofounder Reid learned while at Ford Motor Company that there was a large commercial market for LEO satellite-based PNT services, especially to support autonomous vehicles. Xona Space Systems initially intended to target the commercial market exclusively but soon realized the potential of military and other government applications for its PNT services. At this point, Xona expects to focus primarily on commercial markets, but it recognizes that there is clear value to government users and is continuing to engage and support interested parties in regard to potential government applications.

Status: Are They Selling on the Commercial Market?

As of the second quarter of 2022, Xona had raised venture capital to fund additional development and commercialization of its novel PNT technologies.[141] Xona is not currently selling on the commercial market but anticipates it will. Addressing the commercial market and rollout plans, Xona representatives state that

> Xona's total addressable market (TAM) is 2 billion devices per year, approximately 90 percent of which are mobile devices. The initial serviceable available market (SAM) are the sectors that place high value on superior accuracy, security, or availability, which are the unique selling propositions of the Xona service. The number of devices in the SAM is in the hundreds of millions and expected to grow substantially by 2025. The SAM expands as industries such as autonomous vehicles, aerial mobility, augmented reality, and mobile robotics become more widespread this decade. The automotive sector accounts for a majority of this motivated market, and their requirements ultimately drive the performance metrics of the Pulsar system. Specifically, many downstream markets value centimeter-level precision and high-integrity PNT services.

> Xona's go-to-market strategy is similar to that of new GNSS signals, such as the commercial GNSS corrections services that are operational today.

> There are three phases in the commercial PNT service rollout. Starting in 2025, the first phase of service consists of approximately 40 satellites, which provides one-satellite-in-view coverage over the mid-latitudes. This enables (1) a resilient timing service and (2) GNSS enhancement services. The second phase expands the constellation to 70 satellites, and the one-in-view service expands to become fully global. The third phase further

---

[141] Rainbow, 2021.

expands the constellation to 300 satellites, which provides GPS-level satellite visibility and geometry anywhere on earth. This coverage enables Pulsar to operate as an advanced and independent PNT constellation. As market validation for this commercialization approach, Xona has memorandums of understandings in place with some of the largest receiver manufacturers in the world that produce millions of GNSS devices annually.[142]

Challenges

What Challenges Did They Face in Engaging with and Selling to DoD?
No information to report.

How Did They Address These Challenges?
No information to report.

What Is Next for the Company?
Xona is continuing to mature its technology, drawing on the funding it secured to expand Xona laboratory facilities, conduct more demonstration tests, and begin production manufacturing.[143] Xona is raising additional funds to finance the deployment of its phase 1 satellite constellation rollout and beyond. Xona is also engaged with the Federal Communications Commission (FCC), and domestic and international spectrum stakeholders to broadcast PNT signals globally.

Xona notes that "the Xona Pulsar value proposition has resonated with companies in the construction, agriculture, and automotive markets and is validated by memorandums of understanding of significant financial value."[144] Xona intends to build on these successes and establish similar partnerships with GPS user equipment companies in other markets.

Xona is incorporating its technology in traditional GPS equipment in the hope that it will be a valuable supplement to legacy GPS-only systems as a truly global alternative PNT system. In addition, the company will be continuing to test and refine Xona Pulsar by using higher-fidelity user equipment to boost position accuracy and security, testing performance in GPS-jammed environments, and transmitting signal demonstrations in space in 2022. Finally, Xona is focused on making sure people in the DoD and industry know who they are—and know that the company is engaging with interested stakeholders to deliver PNT services that meet and eventually exceed end user demands across many markets.[145]

---

[142] Xona Space Systems information to authors, April 1, 2022.

[143] Rainbow, 2021.

[144] Xona Space Systems information to authors, April 1, 2022.

[145] Interview 39, Interviewee 50, October 27, 2021.

Acquisition Policy Game Design

We document the game design and inputs here in more detail.

Game Design Approach

Based on the literature review, deep dives, and case studies, the team developed two notional policy regimes that incorporated alternative approaches to strengthen various aspects of the CTP. We refer to each specific alternative approach—a change in how some aspect of the CTP operates—as a policy lever. The policy levers that we tested are discussed in detail in Chapter 5, but for purposes of this methodological description, it is sufficient to say that one grouping represented a minimal change to the status quo while the other represented a more significant set of reforms. In both cases, we assumed that the CTP stakeholder organizations and their roles, responsibilities, authorities, and basic processes (requirements, acquisition, and budgeting) were not fundamentally changed from the status quo. In other words, both sets of reforms would be implemented in a policy and organizational environment substantially the same as exists today.

Based on our research about both the current system and potential reforms under consideration, we developed a game system that represents the major elements of the commercial technology pipeline, which is described in detail in Chapter 3. The system needed to have sufficient complexity that it could credibly mimic the dynamics of the real-world system yet be simple enough that it was both logistically and analytically tractable to manage the moving parts in the game context. Ultimately, we developed a system with five key parts:

1. actors (players) representing specific kinds of CTP stakeholder organizations
2. a policymaking context, including notional strategic objectives and oversight guidance to govern player behavior (rules of the game)
3. a set of notional technologies that players could opt to invest in
4. a set of CTP activities that players use to move a technology through the pipeline
5. a simple resourcing system that provided players constraints on their decisions.

Within this game system, we selected policy levers to test, drawing from the longer list of alternative approaches that we had developed based on the literature review, interviews, and our analysis. We selected policy levers that we assessed would strengthen the CTP, that the Office of the Secretary of Defense has authority to implement (only one, a flexible funding account, would require Congressional approval), that we deemed plausible, and that could be represented in the game. For purposes of the event, we defined success—the desired outcome of the game—as the number and rate of commercial technologies that were incorporated into existing or new defense programs. The resulting policy game was run virtually on Zoom over the course of two half-day sessions in March 2022.[1]

Both days had the same set of players who each played the same role on both days. Players were drawn from three communities: (1) civilian representatives of the DoD identified during interviews and other stakeholder engagements, (2) RAND SMEs with past professional experience working as or with key CTP stakeholders, and (3) uniformed service members currently serving as RAND military fellows with relevant professional experience. Each player was assigned to represent a role based on their past or current professional experience. The first group of stakeholders represented in the game was DIOs. We selected players with recent past or current experience from organizations that represented different institutional perspectives (e.g., service-specific versus joint) and business models. The second group was made up of "end users" who represented stakeholders who would have the opportunity to use the technology, if procured. The third group played the role of program managers who represented traditional acquisition system stakeholders. Other roles, including the vendors for the technologies, Congress, requirement developers, and service and joint oversight organizations like the OSD/R&E and the OSD/A&S, were represented by the RAND team (white cell) either implicitly in the game construct and materials or explicitly when an active decision was required.

On each of the two days of game play, players received a set of policies and a set of notional technologies that made up the game environment. Both half-day sessions focused on a specific set of policies, which were communicated to players in writing in a form that replicated a DoD policy memorandum. Each memo described a set of priorities, incentives, and processes for players to follow during that half-day session. The policy levers and policy environment on day 1 reflected a slight modification of the real-world status quo; this was our projection of what the policy environment would be if policies in effect or emerging at the time the game was designed and played continued to their logical conclusion. Day 2 included a set of policy levers representing more significant change that created an environment that we hypothesized would be more conducive to identifying, developing, and adopting commercial technology for military use. This included more precise and centralized policy guidance, the availability of flexible funding in addition to the baseline funding, and more explicit oppor-

---

[1] The choice to run in a virtual setting was made early in the project to accommodate changing COVID-19 restrictions. The game was designed from the start to be executed in a virtual environment and benefited from past experience running games virtually, including the one documented in Predd et al., 2021.

tunities to collaborate and share information among stakeholders. These policy levers are described in Chapter 5.

In addition, players received a set of notional technologies (11 different technologies each day) in which players could invest during game play. Technologies mirrored but did not replicate dual-use technologies of potential interest to DoD at the time the game was run. Each technology was described using a consistent format that provided information about the vendor, potential military and commercial use cases, technological maturity, and projected schedule and cost for development. The set of technologies was intentionally designed to vary across these dimensions to provide players with a range of values so that we could understand how their decisions varied given different inputs under the two policy regimes.

Game play on both days followed the same general five steps, though the details of the process, such as who was allowed to speak openly to all participants and how much coordination between players was allowed, varied on day 1 and day 2, according to the policy regime in place that day. In the first step, players were given time to read the input materials (policy guidance and descriptions for an initial eight technologies). Then players were asked to identify which technologies were of interest and to explain their interest. In the third step, the research team asked players to take the funding (in the form of resource tokens) that were allotted to them based on their roles representing a specific organization and use them to invest in technologies of interest.

Players "invested" their tokens to move a technology into one of the CTP functions, which include both defense innovation activities and traditional steps of the acquisition process, as depicted in Figure C.1. The number of chips indicated in the figure represents the "cost" of that activity. In addition to varying players' resourcing, we varied the CTP functions into which they could place technologies. For example, a DIO with a budget of three resource tokens would be able to fund one technology for a proof-of-concept demonstration or three technologies into accelerator programs,[2] but it would not be able to fund a live-fire exercise because while they had sufficient resources, their organization type could not invest in that activity (see Figure C.1 key). In step four, after all investments were made, the white cell used a simple probabilistic system to determine what, if any, impact the activity had on the maturity of the technology: Companies could fail (minus 2 TRL), technologies

---

[2] "Broadly speaking, [commercial accelerators] help ventures define and build their initial products, identify promising customer segments, and secure resources, including capital and employees. More specifically, accelerator programs are programs of limited-duration—lasting about three months—that help cohorts of start-ups with the new venture process. They usually provide a small amount of seed capital, plus working space. They also offer a plethora of networking opportunities, with both peer ventures and mentors, who might be successful entrepreneurs, program graduates, venture capitalists, angel investors, or even corporate executives. Finally, most programs end with a grand event, a 'demo day' where ventures pitch to a large audience of qualified investors." Susan Cohen, "What Do Accelerators Do? Insights from Incubators and Angels," *Innovations: Technology, Governance, Globalization*, Vol. 8, No. 3–4, Summer–Fall 2013, pp. 19–25.

could stay at the same maturity level, or technologies could mature (plus 2 TRL). Once players learned the outcomes of their investments, in step five they were then asked to reflect on their investment choices before repeating the five-step process a second time. In the second move for each day, three additional technologies were introduced; players incorporated these technologies into their decisionmaking, along with the adjudicated results of the initial eight technologies.

As shown in Figure C.1, innovation organizations were allowed to invest in technology to send it to an accelerator or to support concept generation, prototype development, proof-of-concept demonstration, or nontraditional acquisition development. End users were permitted to invest in technology to support proof-of-concept demonstration, live-fire exercise, or nontraditional acquisition development or to field the technology as a residual capability or perform limited fielding. Program managers were permitted to invest in technology to support any of the development and fielding activities. End-user support was required for all development and fielding activities except proof-of-concept demonstration and live-fire exercise.

Technologies matured after each game moved through the CTP activity. The cost of each activity is represented by the tiles in Figure C.1. Pathways through the CTP did not need to be sequential or comprehensive—technologies did not need to move to every activity in every CTP phase in sequential order—but their unique pathway did need to reflect the input information players had on each technology (i.e., maturity, stage of development) and information provided after adjudication. The "cost" of each activity could be paid by two or more players working together; the game time allocated to information sharing offered the opportunities for collaboration.

FIGURE C.1
Activities Available to Players

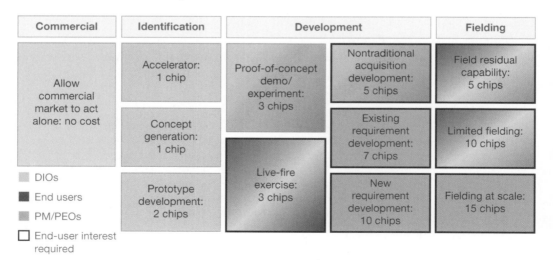

182

Analytic Limitations

The game was designed to support a comparative approach to analysis. Our basic goal was to conduct the game twice over the two days, once modeling each policy regime. Ideally, the only difference between the two runs would have been the policy being tested. In practice, as with most comparative designs,[3] other differences existed that must be accounted for in our analysis. Two are worth noting: (1) the potential for learning between the two runs of the game by using the same participants, and (2) the introduction of new technologies in each round.

For the first issue, we opted to have the same players play through the same basic game twice, once under each set of policy conditions. While this did lead to some risk that player learning between the rounds would confound our findings, in this case, since one policy approach reflected more significant change to the status quo than the other, and since the game process and rules were simple, we felt the potential bias was smaller than the change that would have been introduced between rounds had we introduced a new set of players in the two conditions. In addition, any learning that took place roughly mimics a player gaining experience in the real world.

In the second case, we opted to introduce new sets of technologies at the beginning of each iteration of the game. Because information sharing between different players was a key dynamic in the game, the potential bias caused by player learning would have been substantially larger had we opted to repeat technologies between the two days. To minimize this potential bias, we attempted to develop notional technologies that exhibited a similar range of variation on the two days. However, given the number of factors that could have influenced how players responded to technologies, the two groups of technologies cannot be treated as perfect equivalents.

In addition, there is also the question of how well the game mirrored reality, particularly the number of organizational stakeholders and the number of technologies available for investment.[4] In order to design a four-hour game that would mimic a process that involves dozens of organizations working with hundreds of technologies over years, all aspects of the system were simplified a great deal. This simplification raises questions about how well our findings scale from our simple version of the world. This is particularly salient when it comes to questions about the time and difficulty of making sense of the game environment and coordinating with different actors, since the number of relationships between items scales nonlinearly with increases in the number of items in a system. In other words, doubling the number of actors or the number of technologies represented in the game may have increased the time required to make decisions by more than twice as much, but the game

---

[3] For more on comparative game design methods, see Elizabeth M. Bartels, *Building Better Games for National Security Policy Analysis: Towards a Social Scientific Approach*, Santa Monica, Calif.: RAND Corporation, RGSD-437, 2020.

[4] For social scientists, this will be familiar as the concept of external validity.

does not provide clear bases from which to extrapolate. Thus, while we are fairly confident that the policy levers we implemented would result in improvements to the status quo, we do not know how significant those improvements might be.

In addition, collaboration and information sharing was made easier in the sense that everyone was in the same virtual space, and all game materials (representing information inputs to decisions) were provided to all players. At the same time, game time constraints limited the time available to collaborate and share information in a way that is quite different from the real world. Finally, the personal professional risk any player carried in game decisions was lower than it would be in the real world. For example, in the real world, a PM who reprogrammed funds from one program to support technology development could face negative consequences if the program subsequently failed. This insulation from personal consequences tends to incentivize risk-taking.

Abbreviations

AAL	Army Applications Laboratory
AFRL	Air Force Research Laboratory
ASA(ALT)	Assistant Secretary of the Army for Acquisition, Logistics, and Technology
AUSA	Association of the United States Army
BA	budget activity
BAA	broad agency announcement
C3I	command, control, communications, and intelligence
CCMD	combatant command
CFT	cross-functional team
CNO	chief of naval operations
COTS	commercial off-the-shelf
CP/RRTO	Capability Prototyping/Rapid Reaction Technology Office
CRADA	Cooperative Research and Development Agreement
CRM	customer relationship management
CTO	chief technology officer
CTP	commercial technology pipeline
c-UAS	counter-unmanned aircraft system
DARPA	Defense Advanced Research Projects Agency
DASD(Innovation)	Deputy Assistant Secretary of Defense for Innovation
DAU	Defense Acquisition University
DFAR	Defense Federal Acquisition Regulation
DIE	defense innovation ecosystem
DIO	defense innovation organization
DIU	Defense Innovation Unit
DoD	Department of Defense
DoN	Department of the Navy
ESV	early-stage venture
FAR	Federal Acquisition Regulation
FFRDC	federally funded research and development center
FYDP	Future Years Defense Program
GPS	Global Positioning System
GSA	General Services Administration
ISR	intelligence, surveillance, and reconnaissance

IT	information technology
JEON	Joint Emergent Operational Need
JIEDDO	Joint Improvised Explosive Device Defeat Organization
JRAC	Joint Rapid Acquisition Cell
JUON	Joint Urgent Operational Need
LEO	low earth orbit
MAJCOM	major command
MAPC	Maritime Applied Physics Corporation
NASA	National Aeronautics and Space Administration
NDRI	National Defense Research Institute
NGA	National Geospatial-Intelligence Agency
NPS	Naval Postgraduate School
NSF	National Science Foundation
NSIC	National Security Innovation Capital
NSIN	National Security Innovation Network
O&M	operations and maintenance
OSD	Office of the Secretary of Defense
OSD/A&S	Office of the Secretary of Defense, Acquisition and Sustainment
OSD/R&E	Office of the Secretary of Defense, Research and Engineering
OTA	Other Transaction Authority
OUSD/A&S	Office of the Under Secretary of Defense, Acquisition and Sustainment
OUSD/R&E	Office of the Under Secretary of Defense, Research and Engineering
PEO	program executive office
PIA	partnership intermediary agreement
PM	program manager
PNT	position, navigation, and timing
POM	program objective memorandum
POR	program of record
PPBE	planning, programming, budgeting, and execution
RAA	Rapid Acquisition Authority
R&D	research and development
RDT&E	research, development, test, and evaluation
RF	radio frequency
RFI/RFP	request for information/request for proposal
RNT	regional network team

RRTO	Rapid Reaction Technology Office
SAM.gov	System for Award Management (federal government website)
S&T	science and technology
SBA	U.S. Small Business Administration
SBIR	Small Business Innovation Research
SME	subject-matter expert
SOCOM	special operations command
SOF	special operations forces
STEM	science, technology, engineering, and mathematics
STTR	Small Business Technology Transfer
TRL	technology readiness level
UARC	university-affiliated research center
UAS	unmanned aircraft system
UAV	unmanned aerial vehicle
USAF	U.S. Air Force
USMC	U.S. Marine Corps
USSF	U.S. Space Force
USSOCOM	U.S. Special Operations Command
VC	venture capital

References

AAL—*See* Army Applications Laboratory.

Air Force Research Laboratory, "Tactical Assault Kit," webpage, undated. As of March 16, 2022:
https://afresearchlab.com/technology/information-technology/tactical-assault-kit-tak/

Air Force Research Laboratory Public Affairs, "SpaceWERX Launch Announced During AFWERX Accelerate," December 8, 2020. As of June 8, 2022:
https://www.af.mil/News/Article-Display/Article/2438536/spacewerx-launch-announced-during-afwerx-accelerate/

———, "SpaceWERX Ready to Propel Space Innovation," August 25, 2021. As of June 8, 2022:
https://www.afrl.af.mil/News/Article-Display/Article/2746302/spacewerx-ready-to-propel-space-innovation/

AFWERX, "AFVentures," webpage, undated a. As of January 24, 2022:
https://afwerx.com/afventures-overview/

———, "About Us," webpage, undated b. As of August 3, 2021:
https://afwerx.com/about-us/

———, "Spark," webpage, undated c. As of January 24, 2022:
https://www.afwerx.af.mil/spark.html

———, "Spark Tank," webpage, undated d. As of December 26, 2022:
https://afwerx.com/spark_/spark-tank/

ARCWERX, "Educate," webpage, undated. As of June 1, 2022:
https://arcwerx.dso.mil/educate.html

Army Applications Laboratory, "About the SPARTN Program," webpage, undated a. As of June 13, 2022:
https://aal.army/spartn/

———, "Case Study: The Field Artillery Autonomous Resupply (FAAR) Cohort," Info Sheet, undated b. As of November 8, 2021:
https://aal.army/assets/files/pdf/casestudy-faar-snapshot.pdf

———, "Get Involved," webpage, undated c. As of November 8, 2021:
https://aal.army/get-involved/

———, "SPARTN Info Sheet," undated d. As of June 8, 2022:
https://aal.army/assets/files/pdf/infosheet-spartn.pdf

———, "What We Do," webpage, undated e. As of June 2, 2022:
https://aal.army/what-we-do/

Army Futures Command, "AAL," webpage, undated. As of November 8, 2021:
https://armyfuturescommand.com/aal/

Barber, Ben, "Military's Hummer Shifts to Civilian Market," *The Christian Science Monitor*, January 24, 1994. As of June 14, 2022:
https://www.csmonitor.com/1994/0124/24061.html

Bartels, Elizabeth M., *Building Better Games for National Security Policy Analysis: Towards a Social Scientific Approach*, Santa Monica, Calif.: RAND Corporation, RGSD-437, 2020. As of June 6, 2022:
https://www.rand.org/pubs/rgs_dissertations/RGSD437.html

Bartels, Elizabeth M., Jeffrey A Drezner, and Joel B. Predd, *Building a Broader Evidence Base for Defense Acquisition Policymaking*, Santa Monica, Calif.: RAND Corporation, RR-A202-1, 2020. As of June 6, 2022:
https://www.rand.org/pubs/research_reports/RRA202-1.html

Black Cape, "Focus Areas," website, undated a. As of May 26, 2022:
https://blackcape.io/#focus

———, "Products," website, undated b. As of May 26, 2022:
https://blackcape.io/#products

———, "Rubicon," webpage, undated c. As of February 14, 2022:
https://blackcape.io/products/rubicon

———, "Black Cape, Inc., Announces Emergence from Stealth Mode," press release, December 18, 2019. As of May 26, 2022:
https://www.globenewswire.com/news-release/2019/12/18/1962217/0/en/Black-Cape-Inc
-Announces-Emergence-from-Stealth-Mode.html

Bleicher, Ariel, "Demystifying the Black Box That Is AI," *Scientific American*, August 9, 2017. As of May 26, 2022:
https://www.scientificamerican.com/article/demystifying-the-black-box-that-is-ai/

Both, Thomas, and Nadia Roumani, "Ideation/Creation Expedition Plan: A Flight Plan for Design Exploration," Stanford d.school, webpage, undated. As of January 3, 2022:
https://dschool.stanford.edu/resources/ideation

Budden, Phil, and Fiona Murray, *Defense Innovation Report: Applying MIT's Innovation Ecosystem and Stakeholder Approach to Innovation in Defense on a Country-by-Country Basis*, MIT Lab for Innovation Science and Policy, May 2019. As of June 14, 2022:
https://innovation.mit.edu/assets/Defense-Innovation-Report.pdf

Buesnel, Guy, and Mark Holbrow, "GNSS Threats, Attacks and Simulations," SPIRENT, June 2017. As of March 1, 2022:
gps.gov/governance/advisory/meetings/2017-06/buesnel.pdf

Buhrkuhl, Robert L., "When the Warfighter Needs It Now," *Defense AT&L*, November–December 2006, pp. 28–31.

Business Journals Content Studio, "Vita Inclinata Technologies: winner of the Entrepreneurial Opportunity Contest," *Denver Business Journal*, September 13, 2019. As of October 20, 2021:
https://www.bizjournals.com/denver/news/2019/09/13/vita-inclinata-technologies-finalist.html

Cohen, Susan, "What Do Accelerators Do? Insights from Incubators and Angels," *Innovations: Technology, Governance, Globalization*, Vol. 8, No. 3–4, Summer–Fall 2013, pp. 19–25. As of June 13, 2022:
http://www.mitpressjournals.org/doi/pdf/10.1162/INOV_a_00184

Cordell, Carten, "Dcode Inks 5-Year Contract with Air Force Innovation Hub," *Washington Business Journal*, September 3, 2021. As of May 9, 2022:
https://www.bizjournals.com/washington/news/2021/09/01/dcode-inks-five-year-contract-with
-afwerx.html

Crunchbase, "Compound Eye, Financials," webpage, undated. As of June 6, 2022:
https://www.crunchbase.com/organization/compound-eye/company_financials

DARPA—*See* Defense Advanced Research Projects Agency.

DAU—*See* Defense Acquisition University.

Defense Acquisition University, "Joint Rapid Acquisition Cell," webpage, undated a. As of November. 8, 2021:
https://acqnotes.com/acqnote/acquisitions/joint-rapid-acquisition-cell-jrac

———, "Research Development Test and Evaluation (RDT&E) Funds," webpage, undated b. As of June 6, 2022:
https://www.dau.edu/acquipedia/pages/ArticleContent.aspx?itemid=174

———, "Small Business Innovation," presentation, May 19, 2021. As of May 11, 2022:
https://business.defense.gov/Portals/57/Documents/SB%20%20Innovation%205-19-21.pdf?ver=OtZRuscpzwJ7MClPClbhWQ%3D%3D

Defense Advanced Research Projects Agency, "About DARPA," webpage, undated a. As of April 20, 2021:
https://www.darpa.mil/about-us/about-darpa

———, "DARPA Forward," webpage, undated b. As of June 3, 2022:
https://forward.darpa.mil/

———, "Embedded Entrepreneurship Initiative," webpage, undated c. As of April 20, 2022:
https://eei.darpa.mil

———, "Falcon HTV-2 (Archived)," website, undated d. As of June 20, 2022:
https://www.darpa.mil/program/falcon-htv-2

———, "Hypersonic Air-breathing Weapon Concept (HAWC)," website, undated e. As of June 20, 2022:
https://www.darpa.mil/program/hypersonic-air-breathing-weapon-concept

———, "Industry," webpage, undated f. As of April 20, 2021:
https://www.darpa.mil/work-with-us/for-industry

———, "Office-wide Broad Agency Announcements," website, undated g. As of June 20, 2022:
https://www.darpa.mil/work-with-us/office-wide-broad-agency-announcements

———, "Our Research," webpage, undated h. As of April 20, 2021:
https://www.darpa.mil/our-research

———, "What DARPA Does," webpage, undated i. As of July 21, 2022:
https://www.darpa.mil/about-us/what-darpa-does

———, "Breakthrough Technologies for National Security," March 2015. As of April 20, 2022:
https://www.darpa.mil/attachments/DARPA2015.pdf

———, "Creating Technology Breakthroughs and New Capabilities for National Security," August 2019. As of April 20, 2022:
https://www.darpa.mil/attachments/DARPA-2019-framework.pdf

———, "Small Business Programs Office Transition and Commercialization Support Program Fact Sheet," May 21, 2020. As of April 20, 2022:
https://www.darpa.mil/work-with-us/for-small-businesses/commercialization-continued

Defense Innovation Unit, "About," webpage, undated a. As of June 23, 2022:
https://www.diu.mil/about

——— (experimental), "Commercial Solutions Opening," white paper, undated b.

———, "Defense Innovation Unit (DIU): Who We Are/Our Mission," webpage, undated c. As of June 3, 2022:
https://www.diu.mil/about

———, *Annual Report 2019*, 2019. As of June 9, 2022:
https://assets.ctfassets.net/3nanhbfkr0pc/ZF9fhsMe6jtX15APMLalI/
cd088a59b91857c5146676e879a615bd/DIU_2019_Annual_Report.pdf

———, *Annual Report FY 2021*, 2021. As of October 29, 2022:
https://assets.ctfassets.net/3nanhbfkr0pc/5JPfbtxBv4HLjn8eQKiUW9/
cab09a726c2ad2ed197bdd2df343f385/Digital_Version_-_Final_-_DIU_-_2021_Annual_Report.pdf

Dempsey, Dan, "11 Myths About GPS for Autonomous Vehicles," *Electronic Design*, May 8, 2019,
As of March 1, 2022:
https://www.electronicdesign.com/markets/automotive/article/21807967/11-myths-about-gps
-for-autonomous-vehicles

Dillard, John, and Steve Stark, "Understanding Acquisition: The Valley of Death," United States
Army Acquisition Support Center, October 6, 2021. As of May 5, 2022:
https://asc.army.mil/web/news-understanding-acquisition-the-valley-of-death/

Distributed Spectrum LLC, homepage, undated. As of March 16, 2022:
https://distributedspectrum.com

DIU—*See* Defense Innovation Unit.

DoD—*See* U.S. Department of Defense.

DoD Manual 5000.78, *Rapid Acquisition Authority (RAA)*, Washington D.C.: U.S. Department
of Defense, March 20, 2019.

Doolittle Institute, "Grand Challenge: Metal Additive Manufacturing," webpage, undated a. As
of June 2, 2022:
https://doolittleinstitute.org/event/grand-challenge-metal-additive-manufacturing/

———, "Our History," website, undated b. As of February 14, 2022:
https://doolittleinstitute.org/afrl-innovation-institutes/

———, "AFRL Innovation Institutes," webpage, 2021a. As of November 5, 2021:
https://doolittleinstitute.org/afrl-innovation-institutes/

———, "Technology Transfer," webpage, 2021b. As of November 5, 2021:
https://doolittleinstitute.org/technology-transfer/

Drezner, Jeffrey A., Irv Blickstein, Raj Raman, Megan McKernan, Monica Hertzman, Melissa A.
Bradley, Dikla Gavrieli, and Brent Eastwood, *Measuring the Statutory and Regulatory Constraints
on Department of Defense Acquisition: An Empirical Analysis*, Santa Monica, Calif.: RAND
Corporation, MG-569-OSD, 2007. As of August 2, 2022:
https://www.rand.org/pubs/monographs/MG569.html

Eckstein, Megan, "NavalX Innovation Support Office Opening 5 Regional 'Tech Bridge' Hubs,"
USNI News, September 3, 2019. As of June 10, 2022:
https://news.usni.org/2019/09/03/navalx-innovation-support-office-opening-5-regional-tech
-bridge-hubs

———, "Navy Now Has 15 Tech Bridges Across U.S. and in U.K. to Tackle Fleet Problems in
New Ways," *USNI News*, December 14, 2020. As of June 10, 2022:
https://news.usni.org/2020/12/14/navy-now-has-15-tech-bridges-across-u-s-and-in-u-k-to-tackle
-fleet-problems-in-new-ways

European Union Agency for the Space Programme, "What Is GNSS?" webpage, undated. As of
February 2, 2022:
https://www.euspa.europa.eu/european-space/eu-space-programme/what-gnss

FedTech, "Accelerators," webpage, undated. As of June 6, 2022:
https://www.fedtech.io/accelerators

Fortune Business Insights, "Big Data Analytics Market: 2021 Size, Growth Insights, Share, COVID-19 Impact, Emerging Technologies, Key Players, Competitive Landscape, Regional and Global Forecast to 2028," *GlobeNewswire*, December 16, 2021. As of May 30, 2022:
https://www.globenewswire.com/news-release/2021/12/16/2353210/0/en/Big-Data-Analytics
-Market-2021-Size-Growth-Insights-Share-COVID-19-Impact-Emerging-Technologies-Key
-Players-Competitive-Landscape-Regional-and-Global-Forecast-to-2028.html

Foust, Jeff, "Britain Charts a New Course for Satellite Navigation," *SpaceNews*, August 24, 2021. As of December 8, 2021:
https://spacenews.com/britain-charts-a-new-course-for-satellite-navigation/

Freedberg, Sydney J., Jr., "Hicks Seeks to Unify Service Experiments with New 'Raider' Fund," *Breaking Defense*, June 21, 2021. As of June 1, 2022:
https://breakingdefense.com/2021/06/hicks-seeks-to-unify-service-experiments-with-new
-raider-fund/

Freeman, Jon, Tess Hellgren, Michele Mastroeni, Giacomo Persi Paoli, Kate Cox, and James Black, *Innovation Models: Enabling New Defence Solutions and Enhanced Benefits from Science and Technology*, Santa Monica, Calif.: RAND Corporation, RR-840-MOD, 2015. As of June 13, 2022:
https://www.rand.org/pubs/research_reports/RR840.html

Gallo, Marcy E., *Defense Advanced Research Projects Agency: Overview and Issues for Congress*, Washington, D.C.: Congressional Research Service, R-45088, August 19, 2021. As of June 6, 2022:
https://crsreports.congress.gov/product/details?prodcode=R45088

Garrison, Peter, "Head Skunk," *Air & Space* (Mar. 2010). As of June 13, 2022:
https://www.smithsonianmag.com/air-space-magazine/head-skunk-5960121/

Golden Wiki, "Department of Defense Innovation Ecosystems." As of June, 2, 2022:
https://golden.com/wiki/Department_of_Defense_Innovation_Ecosystems-39ZDW85

Gorski, Eli, "HumanGeo Acquired by Radiant Group," *FCW*, July 27, 2015. As of May 26, 2022:
https://fcw.com/workforce/2015/07/humangeo-acquired-by-radiant-group/247857/

Goward, Dana, "2021 Defense Act Signals Turning Point for Congress and PNT," *GPS World*, December 29, 2020. As of June 6, 2022:
https://www.gpsworld.com/2021-defense-act-signals-turning-point-for-congress-and-pnt/

GPS.gov, "GPS Accuracy," webpage, undated a. As of March 1, 2022:
https://www.gps.gov/systems/gps/performance/accuracy/

———, "GPS Overview," webpage, undated b. As of February 2, 2022:
https://www.gps.gov/systems/gps/#:~:text=The%20Global%20Positioning%20System%20(GPS
,segment%2C%20and%20the%20user%20segment

———, "Space Segment," webpage, undated c. As of February 2, 2022:
https://www.gps.gov/systems/gps/space/

Grissom, Adam R., Caitlin Lee, and Karl P. Mueller, *Innovation in the United States Air Force: Evidence from Six Cases*, Santa Monica, Calif.: RAND Corporation, RR-1207-AF, 2016. As of June 1, 2022:
https://www.rand.org/pubs/research_reports/RR1207.html

Howard, Michael, "xTechSearch 4 Grand Prize Winner's Technology Offers Innovation in Helicopter Rescue Operations," October 22, 2020. As of October 20, 2021:
https://www.army.mil/article/240205/xtechsearch_4_grand_prize_winners_technology_offers
_innovation_in_helicopter_rescue_operations

Inside GNSS, "UK Acquires OneWeb LEO Constellation, but Won't Work for SatNav—or Maybe It Will," July 14, 2020. As of December 8, 2021:
https://insidegnss.com/uk-acquires-oneweb-leo-constellation-but-wont-work-for-satnav/

———, "LEO PNT Service Provider, Equipment Manufacturer Team for Pulsar PNT," November 19, 2021. As of June 6, 2022:
https://insidegnss.com/leo-pnt-service-provider-equipment-manufacturer-team-for-pulsar-pnt/

Insinna, Valeria, "After Hearing Silicon Valley Complaints, Hicks Says No 'Magical' Fix to Acquisition," *Breaking Defense*, April 11, 2022. As of May 2, 2022:
https://breakingdefense.com/2022/04/after-hearing-silicon-valley-complaints-hicks-says-no-magic-bullet-fixing-acquisition/

Joint Rapid Acquisition Cell, "Meeting Warfighter Needs for the Asymmetric Threat," presentation, NDIA Gun and Missile Systems Conference, April 25, 2007. As of June 8, 2022:
https://ndiastorage.blob.core.usgovcloudapi.net/ndia/2007/gun_missile/GMWedGS/ClagettPresentation.pdf

Kirchherr, Julian, and Katrina Charles. "Enhancing the Sample Diversity of Snowball Samples: Recommendations from a Research Project on Anti-Dam Movements in Southeast Asia," *PloS One*, Vol. 13, No. 8, August 22, 2018, e0201710. As of May 31, 2022:
https://www.ncbi.nlm.nih.gov/pmc/articles/PMC6104950/

Landreth, James M., "Through DoD's Valley of Death," Defense Acquisition University, February 1, 2022. As of May 4, 2022:
https://www.dau.edu/library/defense-atl/blog/Valley-of-Death#:~:text=The%20often%20discussed%20%E2%80%9CValley%20of,product%20to%20a%20DoD%20contract

Laurent, Anne, "So Many Innovation Hubs, So Hard to Find Them," *GovExec.com* (September 11, 2019). As of June 2, 2022:
https://www.govexec.com/technology/2019/09/so-many-innovation-hubs-so-hard-find-them/159798/

Leshchinskiy, Brandon, and Andrew Browne, "Digital Transformation Is a Cultural Problem, Not a Technological One," *War on the Rocks*, May 17, 2022. As of June 6, 2022:
https://warontherocks.com/2022/05/digital-transformation-is-a-cultural-problem-not-a-technological-one/

Lewis, James Andrew, *Mapping the National Security Industrial Base: Policy Shaping Issues*, Washington, D.C.: Center for Strategic & International Studies, May 2021. As of June 13, 2022:
https://csis-website-prod.s3.amazonaws.com/s3fs-public/publication/210519_Lewis_NationalSecurity_IndustrialBase.pdf?OToed7f.cbe_LNch7mObAOhx2pLHc21E

LinkedIn, ARCWERX, undated a. As of June 8, 2022:
https://www.linkedin.com/company/arcwerx/about/

———, Compound Eye, undated b. As of June 6, 2022:
https://www.linkedin.com/company/compound-eye

———, Vita Inclinata Technologies, undated c. As of October 20, 2021:
https://www.linkedin.com/company/vitainclinata/about/

Lockheed Martin, "Missions Impossible: The Skunk Works® Story," undated. As of June 13, 2022:
https://www.lockheedmartin.com/en-us/news/features/history/skunk-works.html

Lopez, C. Todd, "Consolidation of Defense Industrial Base Poses Risks to National Security," Defense.gov, February 16, 2022. As of May 26, 2022:
https://www.defense.gov/News/News-Stories/Article/Article/2937898/dod-report-consolidation
-of-defense-industrial-base-poses-risks-to-national-sec/#:~:text=The%20%22defense%20industrial
%20base%22%20refers,needed%20to%20defend%20the%20nation

Lord, Ellen M., and Michael D. Griffin, "Definitions and Requirements for Other Transactions Under Title 10, United States Code, Section 2371 b," memorandum, November 20, 2018. As of May 6, 2022:
https://aaf.dau.edu/wp-content/uploads/2018/11/Definitions-and-Requirements-for-Other
-Transactions-Under-Title-10_USC_S....pdf

Lunden, Ingrid, "What Do We Mean When We Talk About Deep Tech?" *TechCrunch*, March 11, 2020. As of June 6, 2022:
https://techcrunch.com/2020/03/11/what-do-we-mean-when-we-talk-about-deep-tech/

MassChallenge, homepage, undated. as of May 9, 2022:
https://masschallenge.org/

Mayfield, Mandy, "Need for Speed: SOFWERX Zeros In on Rapid Acquisition," *National Defense*, May 10, 2021. As of February 14, 2022:
https://www.nationaldefensemagazine.org/articles/2021/5/10/sofwerx-zeros-in-on-rapid
-acquisition

McQuiston, Barbara, "Statement on Performing the Duties of the Under Secretary of Defense for Research and Engineering," presented to the U.S. Senate, Subcommittee of the Committee on Appropriations, Washington, D.C., April 13, 2021. As of July 21, 2022:
https://www.congress.gov/event/117th-congress/senate-event/LC68180/text?r=2&s=1

McTigue, Kathleen, "Economic Benefits of the Global Positioning System to the U.S. Private Sector Study," NIST, October 2, 2019. As of June 6, 2022:
https://www.nist.gov/news-events/news/2019/10/economic-benefits-global-positioning-system
-us-private-sector-study

Middleton, Michael W., *Assessing the Value of the Joint Rapid Acquisition Cell*, Thesis, Naval Postgraduate School, December 2006.

Miller, Jonathan "jmill," "Bringing Heroes Home, Featuring Caleb Carr of Vita Inclinata," *Tough Tech Today with Meyen and Miller*, November 3, 2020. As of October 20, 2021:
https://toughtechtoday.com/bringing-heroes-home/

MITRE, "DoD Innovation Ecosystem," website, undated. As of June 14, 2022:
https://aida.mitre.org/dod-innovation-ecosystem/

Mosbrucker, Kristin, "New Air Force Innovation Hub in Austin Offers Problem Solving for SA," *San Antonio Business Journal*, December 12, 2018. As of January 24, 2022:
https://www.bizjournals.com/sanantonio/news/2018/12/12/new-air-force-innovation-hub-in
-austin-offers.html

Munn, Zachary, Micah D. J. Peters, Cindy Stern, Catalin Tufanaru, Alexa McArthur, and Edoardo Aromataris, "Systematic Review or Scoping Review? Guidance for Authors When Choosing Between a Systematic or Scoping Review Approach," *BMC Medical Research Methodology*, Vol. 18, No. 143, November 2018, art. 143. As of June 14, 2021:
https://bmcmedresmethodol.biomedcentral.com/articles/10.1186/s12874-018-0611-x

Murphy, Kate, "America Has a GPS Problem," *New York Times*, January 23, 2021. As of June 8, 2022:
https://www.nytimes.com/2021/01/23/opinion/gps-vulnerable-alternatives-navigation-critical
-infrastructure.html

National Academies of Sciences, Engineering, and Medicine, *The Role of Experimentation Campaigns in the Air Force Innovation Life Cycle*, Washington, D.C.: The National Academies Press, 2016. As of November 4, 2022:
https://doi.org/10.17226/23676

———, *Advancing Concepts and Models for Measuring Innovation: Proceedings of a Workshop.* Washington, DC: The National Academies Press, 2017. As of November 4, 2022:
https://doi.org/10.17226/23640

National Guard, "State Partnership Program," webpage, undated. As of June 8, 2022:
https://www.nationalguard.mil/Leadership/Joint-Staff/J-5/International-Affairs-Division/State-Partnership-Program/

National Security Innovation Capital, "About," website, undated. As of October 29, 2022:
https://www.nsic.mil/

National Security Innovation Network, "Emerge Accelerator," webpage, undated a. As of June 10, 2022:
https://www.nsin.mil/emerge/

———, "Foundry," webpage, undated b. As of June 10, 2022:
https://www.nsin.mil/foundry

———, "Hackathons Redefined," webpage, undated c. As of June 6, 2022:
https://www.nsin.mil/hacks-redefined/

———, "Our Regions," webpage, undated d. As of June 3, 2022:
https://www.nsin.mil/regions/

———, "MD5 Adopts New Name to Reflect Refined Mission," webpage, May 6, 2019. As of November 12, 2021:
https://www.nsin.mil/news/2019-05-06-md5-adopts-new-name/

———, "Mission," webpage, 2021a. As of November 2, 2021:
https://www.nsin.mil/mission/

———, "Team," webpage, 2021b. As of November 2, 2021:
https://www.nsin.mil/team/

———"Mad Hacks: Fury Code Concludes with Four Winners," February 26, 2021c. As of June 6, 2022:
https://www.nsin.us/news/2021-02-26-mad-hacks-winners/

———, "NSIN-Program Alumni Received $25,000 for DoD Development," June 3, 2021d. As of June 6, 2022:
https://www.nsin.us/news/2021-06-03-nsin-vector/

Naval Postgraduate School, "NPS Field Experimentation," webpage, undated. As of May 9, 2022:
https://nps.edu/web/fx/what-is-jifx-

NavalX, "Centers for Adaptive Warfighting," webpage, undated a. As of January 18, 2022:
https://www.secnav.navy.mil/agility/Pages/caw.aspx

———, "The NavalX Mission," webpage, undated b. As of October 4, 2021:
https://www.secnav.navy.mil/agility/Pages/default.aspx

———, "Our Team," webpage, undated c. As of January 18, 2022:
https://www.secnav.navy.mil/agility/Pages/team.aspx

———, "Tech Bridges," webpage, undated d. As of January 18, 2022:
https://www.secnav.navy.mil/agility/Pages/tech_bridge.aspx

———, Military Scrum Master Course, Center for Adaptive Warfighting, 2017. As of June 10, 2022:
https://www.secnav.navy.mil/agility/assets/documents/MSM_GUIDE_20201108.pdf

———, "2020 Tech Bridge Annual Report," 2020a. As of October 31, 2022:
https://www.secnav.navy.mil/agility/assets/documents/TB_2020_Annual_Report.pdf

———, "DON Workforce Use of a Tech Bridge," video, YouTube, June 4, 2020b. As of June 10, 2022:
https://youtu.be/4nbyZ3F62sk

Newlab, "NSIN Propel Demo Day," Bryan Chan, Xona Space Systems cofounder presentation,
video, Youtube, September 27, 2021. As of March 2, 2022:
https://www.youtube.com/watch?v=wiiD5jSGQa0&t=3581s

NSIN—*See* National Security Innovation Network.

"NSIN Overview Brief," on file with authors, undated.

Office of the Assistant Secretary of Defense, Acquisition, "Joint Rapid Acquisition Cell,"
webpage, undated. As of November 8, 2021:
https://www.acq.osd.mil/asda/jrac/index.html

Office of the Under Secretary of Defense, Research and Engineering, "Business and Industry,"
website, undated a. As of April 26, 2022:
https://www.ctoinnovation.mil/business-and-industry/#

Office of the Under Secretary of Defense, Research and Engineering, "Innovation Pathways,"
homepage, undated b. As of November 1, 2022:
https://www.ctoinnovation.mil/

OUSD/R&E—*See* Office of the Under Secretary of Defense, Research and Engineering.

Oprihory, Jennifer-Leigh, "ARCWERX Will Leverage ANG, Reserve Innovation to Enhance
Total Force," *Air & Space Forces Magazine*, May 13, 2020. As of June 8, 2022:
https://www.airforcemag.com/arcwerx-will-leverage-ang-reserve-innovation-to-enhance-total
-force/

Predd, Joel B., Jon Schmid, Elizabeth M. Bartels, Jeffrey A. Drezner, Bradley Wilson, Anna Jean
Wirth, and Liam McLane, *Acquiring a Mosaic Force: Issues, Options, and Trade-Offs*, Santa
Monica, Calif.: RAND Corporation, RR-A458-3, 2021. As of June 6, 2022:
https://www.rand.org/pubs/research_reports/RRA458-3.html

Rainbow, Jason, "Xona Space Systems Fully Funds GPS-Alternative Demo Mission," *SpaceNews*,
September 22, 2021. As on March 1, 2022:
https://spacenews.com/xona-space-systems-fully-funds-gps-alternative-demo-mission/

Rapid Reaction Technology Office, "Overview," press release, January 2018. As of May 11, 2022:
https://jteg.ncms.org/wp-content/uploads/2018/02/RRTO-Overview-Brief_Jan2018.pdf

———, "Overview," slide presentation, February 2021. As of May 11, 2022:
https://nps.edu/documents/104517539/126210406/2021+Dist+A+USD%28R%26E%29+RRTO
+Overview+26+Feb+2021.pdf/90637caa-3574-0b8e-0a41-9cf86d6f7afb?t=1614646251193

———, "2021 Global Needs Statement," *SAM.gov*, March 24, 2021. As of May 11, 2022:
https://sam.gov/opp/d1c8253de5ed42c394d858aac1b90427/view?keywords=%22RRTO%20
Global%20Needs%20Statement%22&sort=-relevance&index=opp&is_active=true&page=1

Reid, Tyler Gerald Rene, *Orbital Diversity for Global Navigation Satellite Systems*, Palo Alto
Calif.: Stanford University, June 2017. As of March 2, 2022:
https://web.stanford.edu/group/scpnt/gpslab/pubs/theses/TylerReidThesis2017.pdf

Resilient Navigation and Timing Foundation, "OneWeb PNT Possibilities," blog post, April 19, 2021. As of December 8, 2021:
https://rntfnd.org/2021/04/19/oneweb-pnt-possibilities/

"Rights in Technical Data—Noncommercial Items, 252.227-7013," *Defense Federal Acquisition Regulation Supplement*, February 2014. As of January 4, 2022:
https://www.acq.osd.mil/dpap/dars/dfars/html/current/252227.htm#252.227-7013

RRTO—*See* Rapid Reaction Technology Office.

Rudin, Cynthia, and Joanna Radin, "Why Are We Using Black Box Models in AI When We Don't Need To? A Lesson from an Explainable AI Competition," *Harvard Data Science Review*, Vol. 1, No. 2, November 22, 2019. As of May 26, 2022:
https://hdsr.mitpress.mit.edu/pub/f9kuryi8/release/7

Sam.gov, System for Award Management, webpage, undated. As of July 20, 2022:
https://sam.gov/content/home

Schweikhart, Sharon A., and Allard E. Dembe, "The Applicability of Lean and Six Sigma Techniques to Clinical and Translational Research," *Journal of Investigative Medicine*, Vol. 57, No. 7, October 2009, pp. 748–755.

Section 809 Panel, *Report of the Advisory Panel on Streamlining and Codifying Acquisition Regulations*, Vol. 1, January 2018. As of July 20, 2022:
https://acqnotes.com/wp-content/uploads/2018/07/Section-809-Advisory-Panel-for-Streamlining
-Acquisition-Regulations-Volume-1-Jan-2018.pdf

———, *Report of the Advisory Panel on Streamlining and Codifying Acquisition Regulations*, Vol. 2, June 2018.

———, *Report of the Advisory Panel on Streamlining and Codifying Acquisition Regulations*, Vol. 3, January 2019.

SBIR.gov, "Platform Agnostic Data Storage Infrastructure," webpage, undated a. As of May 26, 2022:
https://www.sbir.gov/node/1962811

———, "STARbase: An Innovative Solution for Platform Agnostic Data Storage," webpage, undated b. As of May 26, 2022:
https://www.sbir.gov/node/1962687

Smith, Giles K., Jeffrey A. Drezner, William C. Martel, James J. Milanese, W. E. Mooz, and E. C. River, *A Preliminary Perspective on Regulatory Activities and Effects in Weapons Acquisition*, Santa Monica, Calif.: RAND Corporation, R-3578-ACQ, 1988. As of August 3, 2022:
https://www.rand.org/pubs/reports/R3578.html

SOFWERX, "About SOFWERX," webpage, undated a. As of June 10, 2022:
https://www.sofwerx.org/about

———, "Creating Opportunities," webpage, undated b. As of June 10, 2022:
https://www.sofwerx.org/impact/

———, "Events: Rapid Prototyping Event," webpage, undated c. As of June 10, 2022:
https://events.sofwerx.org

———, homepage, undated d. As of June 10, 2022:
https://www.sofwerx.org

———, "Join the Mission," webpage, undated e. As of June 10, 2022:
https://www.sofwerx.org/ecosystem

———, "Tech Tuesday," webpage, undated f. As of June 10, 2022:
https://www.sofwerx.org/techtuesday/

Space.com, "Superfast Military Aircraft Hit Mach 20 Before Ocean Crash, DARPA Says,"
August 18, 2011. As of June 20, 2022:
https://www.space.com/12670-superfast-hypersonic-military-aircraft-darpa-htv2.html

SPACEWERX, "About Us," webpage, undated. As of June 8, 2021:
https://spacewerx.us/about-us/

TechLink, "Air Force Research Laboratory Munitions Directorate," webpage, undated. As of
October 31, 2022:
https://techlinkcenter.org/labs/afrl-rw/89eb163a-63a0-499e-a069-a2d46f51c47e

———, *National Economic Impacts from the DoD SBIR/STTR Program, 1995–2018*, October 3,
2019. As of June 9, 2022:
https://techlinkcenter.org/economic-impact-reports/national-economic-impacts-of-department
-of-defense-sbir-sttr-program-1995-2018

Temin, Tom, "Who Says Small Innovators Can't Get Big Federal Contracts," *Federal News
Network*, May 10, 2022. As of June 8, 2022:
https://federalnewsnetwork.com/commentary/2022/05/who-says-small-innovators-cant-get
-big-federal-contracts/

Tompkins, Stefanie, "Statement by Dr. Stefanie Tompkins, Director Defense Advanced
Research Projects Agency (DARPA)," submitted to the U.S. Senate Armed Services Committee,
Subcommittee on Emerging Threats and Capabilities, April 6, 2022. As of December 5, 2022:
https://www.armed-services.senate.gov/imo/media/doc/PASSBACK%20DARPA_%20Tompkins
%20SASC-ETC%20testimony%206%20Apr%202022_DARPA_FIINAL%200031.pdf

Ulrey, Scott, and Diane Sidebottom, "Other Transactions Authority," DARPA Contracts
Management Office, presentation, May 2022. As of November 18, 2022:
https://acquisitioninnovation.darpa.mil/docs/Training/Other%20Transactions%20Comprehensive
_May%202022.pptx

U.S. Air Force, *DoD FY 2023 Budget Estimates, Air Force Justification Book*, April 2022. As of
November 3, 2022:
https://www.saffm.hq.af.mil/Portals/84/documents/FY23/PROCUREMENT_/FY23%20Air
%20Force%20Other%20Procurement.pdf?ver=4WRTyibMkeq034KgK44LHg%3D%3D

U.S. Air Force Public Affairs, "Air Force Opens Doors to Universities, Small Businesses and
Entrepreneurs to Boost Innovation," July 21, 2017. As of November 3, 2021:
https://www.af.mil/News/Article-Display/Article/1254932/air-force-opens-doors-to-universities
-small-businesses-and-entrepreneurs-to-boo/

U.S. Army, "DEVCOM Ground Vehicle System Center," webpage, undated. As of June 6, 2022:
https://www.usarmygvsc.com/

U.S. Department of Defense, "MD5—A New Department of Defense National Security
Technology Accelerator—Officially Launches with Disaster Relief Hackathon in New York
City," press release, October 14, 2016. As of November 12, 2021:
https://www.defense.gov/News/Releases/Release/Article/974626/md5-a-new-department-of
-defense-national-security-technology-accelerator-offici/

———, *Summary of the 2018 Department of Defense Artificial Intelligence Strategy*, Washington D.C.,
February 12, 2019. As of June 6, 2022:
https://media.defense.gov/2019/Feb/12/2002088963/-1/-1/1/SUMMARY-OF-DOD-AI
-STRATEGY.PDF

——, *Fiscal Year (FY) 2022 Budget Estimates: Army Justification Book Volume 3a of 3*, May 2021. As of November 8, 2021:
https://www.asafm.army.mil/Portals/72/Documents/BudgetMaterial/2022/Base%20Budget/rdte/RDTE_BA_6_FY_2022_PB.pdf

——, *Fiscal Year (FY) 2022 Budget Estimates: Defense-Wide Justification Book Volume 3 of 5*, May 2021. As of November 2, 2021:
https://comptroller.defense.gov/Portals/45/Documents/defbudget/fy2022/budget_justification/pdfs/03_RDT_and_E/RDTE_Vol3_OSD_RDTE_PB22_Justification_Book.pdf

——, DoD Instruction 5000.85, "Major Capability Acquisition," November 4, 2021. As of November 18, 2022:
https://www.esd.whs.mil/Portals/54/Documents/DD/issuances/dodi/500085p.pdf?ver=2020-08-06-151441-153

——, "Defense Department New Website to Navigate Innovation Opportunities," press release, April 22, 2022. As of April 26, 2022:
https://www.defense.gov/News/Releases/Release/Article/3008046/defense-department-new-website-to-navigate-innovation-opportunities/

——, DoD Instruction 5000.02, "Operation of the Adaptive Acquisition Framework," June 8, 2022. As of November 18, 2022:
https://www.esd.whs.mil/Portals/54/Documents/DD/issuances/dodi/500002p.pdf?ver=2020-01-23-144114-093

U.S. Government Accountability Office, *Defense Science and Technology: Adopting Best Practices Can Improve Innovation Investments and Management*. Washington, D.C.: GPO, GAO-17-499, 2017.

U.S. Small Business Administration, SBIR/STTR America's Seed Fund, "Frequently Asked Questions," webpage, undated a. As of June 6, 2022:
https://www.sbir.gov/faq/vc-participation

——, "The SBIR and STTR Programs," webpage, undated b. As of July 21, 2022:
https://www.sbir.gov/about

——, "Tibbets and Hall of Fame," webpage, undated c. As of October 2, 2021:
https://www.sbir.gov/about-tibbetts-awards

——, "VC Ownership Authority," webpage, undated d. As of June 6, 2022:
https://www.sbir.gov/vc-ownership-authority

Vavasseur, Xavier, "U.S. Navy Successfully Tests GARC/TALONS for LCS MCM Mission Package," *NavalNews.com*, November 26, 2019. As of October 22, 2022:
https://www.navalnews.com/naval-news/2019/11/us-navy-successfully-tests-garc-talons-for-lcs-mcm-mission-package/

Vergun, David, "Official Says DOD, with Help from Partners, on Cusp of Cutting-Edge Innovations," *Defense.gov*, November 8, 2021. As of June 6, 2022:
https://www.defense.gov/News/News-Stories/Article/Article/2837650/official-says-dod-with-help-from-partners-on-cusp-of-cutting-edge-innovations/

Vincent, Brandi, "Space Force's Innovation Hub Announces Solicitations in the Works," *Nextgov.com*, August 12, 2021. As of November 24, 2022:
https://www.nextgov.com/emerging-tech/2021/08/space-forces-innovation-hub-announces-solicitations-works/184487/

Vita Inclinata, "About: Building Technology That Brings People Home Save Every Time," webpage, undated a. As of October 20, 2021:
https://vitatech.co/about/

———, "Caleb Carr," webpage, undated b. As of October 20, 2021:
https://vitatech.co/caleb-carr-ceo/

———, "Our Team," webpage, undated c. As of October 20, 2021:
https://vitatech.co/team/

———, "Vita's Systems: LSS Technology Products," webpage, undated d. As of October 20, 2021:
https://vitatech.co/products/

———, "Vita's Systems: LSS Technology Products: Hoist Rescue," webpage, undated e. As of October 20, 2021:
https://vitatech.co/products-hoistrescue/

———, "Vita's Systems: LSS Technology Products: Litter Attachment," webpage, undated f. As of October 20, 2021:
https://vitatech.co/sling-load/

———, "Vita's Systems: LSS Technology Products: Load Navigator," webpage, undated g. As of October 20, 2021:
https://vitatech.co/products-navigator/

———, "Vita's Systems: LSS Technology Products: Load Pilot," webpage, undated h. As of October 20, 2021:
https://vitatech.co/products-loadpilot/

———, "Vita's Systems: LSS Technology Products: Precision Rapid Aerial Employment System (PRAES)," webpage, undated i. As of October 20, 2021:
https://vitatech.co/praes/

Wickr, "Wickr Selected as Only Secure Communication Platform for Strategic Expansion Initiative with U.S. Air Force," press release, April 9, 2020. As of June 8, 2022:
https://wickr.com/secure-communication-platform-for-air-force/

Wilkinson, David, "Innovation Capacity: How to Develop It in Your Organization," *The OR Briefings*, blog, undated. As of June 8, 2022:
https://oxford-review.com/developing-innovation-capacity/

Xona Space Systems, "About," webpage, undated a. As of March 2, 2022:
https://www.xonaspace.com/about

———, "Background," webpage, undated b. As of June 6, 2022:
https://apps.fcc.gov/els/GetAtt.html?id=279081&x=

———, "Xona Pulsar," webpage, undated c. As of June 6, 2022:
https://www.xonaspace.com/pulsar

———, "Xona Space Systems Raises $1M Pre-Seed Round from 1517, Seraphim Capital, Trucks Venture Capital and Stellar Solutions," press release, May 14, 2020. As of February 2, 2022:
https://www.xonaspace.com/pr20200514

———, "Xona Space Systems Awarded Competitive Grant from the National Science Foundation," press release, December 21, 2020. As of April 1, 2022:
https://www.xonaspace.com/amsim

xTech, "Accelerator," webpage, undated a. As of November 8, 2021:
https://www.arl.army.mil/xtechsearch/accelerator.html

———, "xTechSearch Background," webpage, undated b. As of November 27, 2021:
https://www.arl.army.mil/xtechsearch/xtechsearch.html

———, "xTechSearch 4," webpage, undated c. As of October 20, 2021:
https://www.arl.army.mil/xtechsearch/competitions/xtechsearch-4.html#judging_criteria

———, "xTechSearch 5 Finalist Technology Overview: Compound Eye," video, YouTube, September 16, 2021. As of June 6, 2022:
https://youtu.be/lFOsFeZW44w

Zangvil, Yoav, "Research on GPS Resiliency and Spoofing Mitigation Techniques Across Applications," presentation, undated. As of March 1, 2022:
https://www.gps.gov/governance/advisory/meetings/2019-06/zangvil.pdf